REALISM AND THE CLIMATE CRISIS

Hope for Life

John Foster

First published in Great Britain in 2022 by

Bristol University Press
University of Bristol
1–9 Old Park Hill
Bristol
BS2 8BB
UK
t: +44 (0)117 954 5940
e: bup-info@bristol.ac.uk

Details of international sales and distribution partners are available at bristoluniversitypress.co.uk

British Library Cataloguing in Publication Data
A catalogue record for this book is available from the British Library

ISBN 978-1-5292-2326-2 hardcover
ISBN 978-1-5292-2327-9 paperback
ISBN 978-1-5292-2328-6 ePub
ISBN 978-1-5292-2329-3 ePdf

Cover design by Clifford Hayes
Front cover image: istock.com – Sjo
Bristol University Press uses environmentally responsible print partners.
Printed and bound in Great Britain by CMP, Poole

For
Jean Mary Foster
(1923–2019)
and
Oaklan Foster Pickup
(b. 2017)

Contents

A Note on Notes

The text following has been left deliberately undisfigured by a rash of bracketed references and little superscript numbers. The inclusion of these, which I have always considered a nuisance in any context, becomes when one is aiming at a wider-than-academic audience a positive distraction from the flow of the argument. Wherever an acknowledgement is appropriate or a citation called for, these can still be found, by those interested, in the Notes at the end of the book, which are there identified to the chapter, and within these to the pages and parts of the text, to which they relate.

Acknowledgements

This book tries to improve on my *After Sustainability*, which in its turn tried to improve on *The Sustainability Mirage*. While this is a process which could in principle run and run, I am luckily now old enough for it to be nearing its natural end.

After Sustainability started from a challenge to widely prevalent denial of climate and environmental tragedy, and the hopes with which it concluded were correspondingly both chastened and limited. I have since come to see that one must start from more robustly transformative hopes, while still recognizing that our tragic situation conditions the actions which can flow from them. This shift of emphasis was prompted partly by the personal considerations mentioned briefly in the Introduction, but friendly and constructive criticism of the earlier approach from a good many people also helped; I must especially thank in this connection Samantha Earle, Margaret Gearty, Susan Goldsworthy, Steve Gough, Doris Hauser, Stefan Morales and Rupert Read.

Thanks are due, too, to my Lancaster colleague Matthew Johnson, general editor of the journal *Global Discourse*, for inviting me to guest-edit a special issue (Volume 7, Issue 1, 2017) on the theme 'After Sustainability – What?' – later published in book form as *Post-Sustainability: Tragedy and Transformation* (Abingdon: Routledge, 2018). This project provided a forum for illuminating engagement with a number of my fellow contributors, notably Ingolfur Blühdorn, Mike Hannis and Panu Pihkala. I am particularly grateful to Rupert Read, my exchange with whom extending over the last 40 pages of this volume was a very powerful stimulus to rethinking.

Similarly, involvement with the think tank Green House, including editing its most recent book *Facing Up to Climate Reality: Honesty, Disaster and Hope* (London: Green House, 2019) and contributing a series of opinion pieces and book reviews to its website at https://www.greenhousethinktank.org/, has brought me many valuable insights from colleagues: thanks in particular to Nadine Andrews, Anne Chapman, Ray Cunningham, Jonathan Essex and Brian Heatley. (Neither they nor Green House, however, should be taken to agree with all the views expressed in this book.)

I have been honoured with opportunities to explore and develop my thinking on all these matters in seminars and presentations given at venues including Ashridge Business School, the Vienna University of Economics and Business and the universities of Anglia Ruskin, Bath, Cumbria, Dundee, East Anglia, Lancaster and Surrey. I am very grateful for all the ideas, objections, suggestions and (occasional) plaudits which these discussions have afforded.

I have to thank three publisher's referees, in two cases for supportively helpful comments and in the other for stimulating provocation.

Finally and on a different note, thanks once again to my wife Rose for her loving patience and support. She too will not agree with everything here, but she knows that one has to speak one's truth as best one can.

Introduction: Hope, Realism and the Climate Crisis

> to hope till Hope creates
> From its own wreck the thing it contemplates
>
> Shelley, *Prometheus Unbound*

The possibility of hope is now the central question of our time. That is because it is crucial to the climate crisis, which is our time's overwhelmingly urgent challenge.

Much else is pressing: poverty, hunger, war and threats of war, unravelling international institutions, clashing religious fundamentalisms, cyber-security, the dark web, deep uncertainties around sexuality and identity … the list goes on. This is altogether the most existentially exacting juncture in human history. But the climate crisis is now absolutely primary. On how we respond to that crisis depends, it is increasingly apparent, the future of our existence itself. There can no longer be any serious doubt that the present trajectory of human-induced global heating, unaddressed or even just inadequately addressed, could within the present century take the Earth's atmosphere to temperatures at which civilization certainly, and maybe human life, could not survive. That claim is so far from being irresponsibly alarmist that it expresses the current sober consensus among informed scientists. It would be excessive to say that in such a context, nothing else matters. But the basic preconditions for anything else to go on mattering all that much for all that long are now at risk. When your house is on fire, first things first is a maxim of mere common sense.

With a house on fire, however, acting hopefully on that maxim is likely to be a fairly simple matter. You will hope for some swiftly available means of putting the fire out, and often this hope will be a perfectly realistic one to entertain in your circumstances, and will be duly answered. With the climate crisis, however, things are a lot less straightforward.

Grounds for hope?

In the first place, what we must hope for seems to escape ever further beyond our reach. Every day the odds against humanity's getting itself out of its current mess lengthen – more coal is mined and oil or gas extracted, more carbon dioxide is poured into the atmosphere, more plastic accumulates in the oceans, sea levels slowly but inexorably rise, habitat disruption threatens more species with extinction as registered in each successive scientific report ... and meanwhile, there seems even yet no real appetite among governments, corporations or the public at large to do more than make large declaratory statements regretting these processes, at the same time persisting with something very much like business as usual. Cold-eyed consideration of these conditions suggests that we shall soon be, if we are not already, on an unalterable course towards ecological catastrophe. For, secondly, the scale and scope of even the minimum changes necessary for avoiding such catastrophe make the difficulties of achieving them seem almost insurmountable. These changes include, at least: phasing out fossil fuels and moving to completely carbon-neutral energy generation; ending the apparently irresistible drive to urbanization, both because cities are energy-intensive and because many of them are vulnerable to inevitably rising sea levels; reversing globalization in favour of localized economic resilience (and thus vigorously subverting presently existing capitalism); and starting to apply the precautionary principle as a routine default planning strategy across the board – not least, where increasingly tempting technological quick-fixes for anthropogenic climate change itself are in question. In other words, we are going to require massive and epochal shifts in our established patterns of economic, cultural and political life. And those shifts will have to be worldwide. Even if the prospects for them were vastly less challenging than they already are in the democracies of Western Europe and (just about imaginably) the United States, achieving them there would be effectively useless in terms of halting or reversing global heating unless the rapidly expanding coal-based economies of China and India, at least, could be brought to follow suit. (These economies between them accounted for most of the 70 per cent increase in global demand for coal between 2000 and 2010, and although this growth has slowed in more recent years, there are signs that China in particular may be planning further major expansion in coal-fired capacity.) Finally, despite our having known about the broad parameters of this peril for some five decades, the necessary changes now have to be put in place within no more than the twenty years which we maybe have left before the trajectory towards an atmospheric temperature increase threatening the planet's habitability becomes irresistible.

It is no surprise, then, that Extinction Rebellion starts from the claim that we are facing an unprecedented global emergency, nor that the school strike movement announces its intention 'to rise up and take direct action where older generations have failed'. These closely linked recent manifestations, responding to urgently perceived danger, have energized a wider green constituency which was becoming stale and exhausted from the sheer effort of having tried for years to make headway in pushing this enormous boulder up this steepest of hills. That is by no means to dismiss the established political, NGO and academic strands of green activity, which remain importantly in play. But the whole issue has been decisively reinvigorated and has claimed a new level of public attention through the activities of these newcomers. They are easily the most hopeful gleams of light across what is nevertheless a rapidly darkening landscape. But equally, and just as such, they raise what I have called the question of the possibility of hope most acutely.

That question, put starkly, is: how in these conditions can hope be kept *honest*?

Of course, there is hope and hope. When Greta Thunberg says 'I don't want your hope. I don't want you to be hopeful. I want you to panic. … And then I want you to act', she is calling out the 'hopefulness' paraded at jamborees like the World Economic Forum, a smart-suited glibness substituting for serious action and shrugging off real responsibility. That attitude can't be *kept* honest, because it never was. But climate activists, Ms Thunberg included, have to start from genuinely robust hope, literally morning by morning. They know that in order to carry on, they must tap daily into the energy that only comes from hoping – from always-renewed commitment to the achievability of a sustainably habitable world. And for anyone seriously worried about these issues, activist or not, such commitment is a condition of remaining constructively engaged, rather than turning away to the many much more tractable concerns which life (or at any rate, life in the short term) still offers. Nor is starting afresh, again and again, from such hope merely a therapeutic resource, a kind of moral exercise routine to make activism or constructive engagement easier to maintain. It is what makes them feasible at all, for anyone even minimally attentive to the situation as it now stands. And for hope of that order, the question of honesty does inescapably arise.

Importantly, *starting* from genuine hope means just that, rather than reaching it as the last hypothesis standing when optimism fails. My own work in this field was initially, back in the 1990s, probably an exercise in such optimism: an attempt to defend the then-new sustainability imperative from perverse economistic modelling, as it started to be taken up into the mainstream. There appeared, back then, to be at least some prospect that something better than 'sustainable development', with its quantified and monetized bottom lines, could become the pragmatic governing model. But thereafter I moved, at first very reluctantly but with (it soon felt) no responsible alternative, towards pointing out the danger of our being betrayed by this picture as by a mirage. Sustainable development benchmarks and targets were so readily mainstreamed, it seemed to me, precisely because they served as constraints from the future which never, when push came to shove, actually constrained present activities – rather they provided, in their socially-constructed flexibility, a goal which would always recede before us as we offered to pursue it. It was no long step from identifying that danger to recognizing that having *been* so misled, we had missed the crucial window of opportunity for preventing climate-driven disaster – we simply had not done what we needed to have started doing thirty years previously if we were going to interrupt that trajectory. By that point, hope seemed to be all you found yourself left with when, refusing any longer to blank out the grim prospects awaiting us, you still somehow didn't give up. Hence the *Denial* and *Hope* included in the subtitle of my last book, *After Sustainability*.

But active hope cannot really work as this kind of fallback option. Actually, it has to be our first premise, when we confront climate and environmental crisis threatening life, because it lies at the very roots of the life that is threatened. I recognized this with a painful jolt when recently becoming a grandparent for the first time. For that first grandson, born in the autumn of 2017, will reach my present age in 2088, and on current trends the world will by then be experiencing all the consequences of a 4°C rise in global atmospheric temperature – the Arctic ice cap gone, sea levels risen by a metre or more, large swathes of the tropics uninhabitable or agriculturally useless and corresponding geopolitical turmoil beside which our present difficulties with migration will look like a picnic. 'Future generations', in other words – those whom in earlier phases of green activism we had to project – are already here. One can hold them in one's arms, and acute anxiety about what awaits them becomes direct, chagrined consciousness of how humanly intolerable is what we are preparing for these innocents by what we are doing to the world. Only the upwelling hope which comes naturally and irresistibly along with fresh life itself – *life-hope*, as one may call it – will do as a response.

Hope, in other words, is not optional in face of the climate emergency. But hope has also to be *realistic*. That is a conceptual requirement, an essential part of what the very idea of hope contains. Hope is a more complex attitude to the future than merely wanting something desirable to come about. It is future-directed desire under the sign of contingency. As pure life-energy surging up in us, indeed, it need not take an intentional object at all: we can be, simply, hopeful (this undirected energy often informs great art – think of Rembrandt's painting, say, or Beethoven's music), and thus far, ordinary contingency has not arisen. But as soon as we bring that disposition to bear on practical life, we do so in terms of some goal or project which for the time being expresses it. And then to hope for X is to want X to happen, acknowledging that it might not but at the same time believing that it genuinely might. We may call the latter requirement the condition of realism – genuine hope must address itself to a real chance, however slender, of the hoped-for thing's coming to pass, just as it must also involve the courage to run the risk that it won't. But this conceptual requirement reflects our recognition of a vital practical necessity. Crucially, aspirations which fail to meet the condition of realism can't do the hard work of hope – the work of getting us out of bed each day to go on risking ourselves in what might be a struggle against the grain – because they always turn out to be about what someone can or can't bear to contemplate, rather than about grappling with the tough and recalcitrant world over against us: and one way or another, we always know this. Hoping unrealistically is always either a denial of the full difficulties confronting us, or a way of giving up while pretending not to.

That is why the question is one of honesty, and we can now formulate it more sharply. Can hope for escape from humanity's climate plight any longer count as realistic? Do the odds – which, as I began by sketching them, are now stacked so high against us – leave us anything which could legitimately be claimed as a real chance of averting runaway global overheating? For if they don't, then since life-hope seems non-optional, it can only be embraced as a form of self-delusion. This fear must gnaw at the heart of any climate activist who allows him- or herself to think dispassionately about the issues.

Tough hope

I want to argue that hope can still meet the condition of realism. But it can do so only under other conditions which many activists may find almost as difficult to accept as hope's abandonment. That is why, within its class – books about the climate crisis which contend that we might still have a chance to survive it – this book will be found to take an unconventional line.

For one thing, although it has been substantially revised and brought to a conclusion during the ravages of the COVID-19 pandemic, it does not venture the claim that trying to deal with that grisly outbreak has better prepared us for confronting the climate emergency. It points neither to government willingness to upend the economy, public acceptance of intrusive regulation, nor a new sense of communal solidarity, as potential COVID-19 bonuses to be carried forward into the climate context. For reasons about which I say a little more in the final chapter – essentially, the difference between an immediate mortal threat and prospective disasters well over the horizon – I am seriously sceptical of that claim, and would remain so even were public acceptance of restriction much less patchy than it has latterly become.

But even more unexpectedly, perhaps, the book also includes no material on climate justice (except in passing as part of a critique of the Green New

Deal), and treats neither of human rights nor of alleged rights of any other kind. It does not discuss techno-fixes, whether favourably or unfavourably. It lacks chapters on corporate malfeasance and the sins of late capitalism, the Millennium Development Goals and the scandal of world poverty, the role of green business and investment, and even the prospects for China. It lauds neither the virtues of democracy despite its failures, nor the wisdom of indigenous peoples despite their regrettable marginalization. And it scarcely alludes to issues of Western diet.

These omissions do not mean that I think none of the above matter. They most of them clearly do, very much – our attitudes to technology and to democracy in particular. But they represent a terrain which has now been very well worked over in that sort of book during the past decade or so, from Tim Jackson's *Prosperity Without Growth* through Naomi Klein's *This Changes Everything* and George Monbiot's *Out of the Wreckage* to Jonathon Porritt's recent *Hope in Hell*. And the working-over has answered to a general approach: 'Surely now at last we are reaching a moment when all these progressive concerns must jump together, and serious action to avert climate catastrophe must garner widespread popular consent!' – an approach which has persisted largely unaltered, it has to be said, while the jumping together, and the widespread consent, have (broadly) continued not to materialize.

All these authors, that is, want to find or renew hope *on liberal-progressivist terms*. They want us to be able to hope, even yet, for a transformation which will yield conditions of general material sufficiency, structured by justice and organized by egalitarian-democratic methods. They want to retain, even in this extremity and however sobered and muted, essentially the Enlightenment package.

Instead of revisiting this terrain or maintaining this approach, I want to make an uncomfortably different case – one which I believe we cannot any longer postpone considering. Given the conceptual relations between hope, experience, realism and transformation, in the context of our wholly unprecedented plight, hope can now only be honestly entertained *on tragic terms*. What that entails in practice is that we have to try to save the habitability of the planet *at whatever cost*. And politically, that in turn involves accepting that nothing decisive will happen, on any of these fronts, until the minority who are both intelligent and imaginative enough to know what is coming, and at the same time honest and brave enough to reject complicity in biocide, finally takes matters into its own hands – by whatever means offer themselves, and without waiting for popular consent. Nothing will really happen, that is, until people forming what I call the life-responsible vanguard creatively reconstitute themselves as a new kind of revolutionary force, something markedly different from being a demonstrator or even a 'rebel'.

That, incidentally, is why the book also does not touch on the recently popular concept of 'transformative adaptation', or at least not as such. I think this coinage is a way of saying *revolution* without quite saying it – something which it can manage because, plainly enough, any transformation in patterns of working, habits, expectations and general arrangements likely to be adequate to climate emergency will necessarily go far enough to effect a wholesale turnaround (that is, a 'revolution') in present living; but whether it must involve 'revolutionary' political methods (including deliberate subversion of established state institutions by a combative minority) can then seem to be a further question. For reasons which I canvass most extensively in Chapter 7, however, I believe that change of the order needed will have to be revolutionary in *both* senses, and I prefer to be frank about this from the outset.

In offering to develop such a case, however, I am painfully aware of the danger of being driven in that direction too much by sheer impatience. When, for instance, David Attenborough concludes his impressive and authoritative recent testimony on these matters with the hope that 'brilliant minds' may now be coming together worldwide to understand our problems and at last find solutions, might it not be a mistake for someone who has been aware of the climate and ecological threat for decades to react dismissively? Attenborough insists that 'we can yet make amends, manage our impact, change the direction of our development ...' – all we require, he claims, is the will to go on urging our politicians and to keep faith with 'the labours of countless committees and conferences, and the signing of innumerable international treaties'. Shouldn't we perhaps see this, not as a reiteration of failed conventionalities for which it is now far too late, but as a sign that a powerful Establishment is finally seeing the light? Sir David, with all his grounded wisdom, is a quintessentially Establishment figure, and that someone in that position should be canvassing radical proposals which fringe Greens were advocating thirty years ago could perhaps be welcomed as the long-delayed cutting-through of that advocacy – and maybe still, after all, just in time? Would it not be a betrayal to turn aside, either in frustration or in a too-ready contempt for human folly, from that trodden route through our existing politics and institutions, at the very moment when it begins to have a chance of taking us somewhere?

That question puts one, quite rightly, on the spot. I am led to the answer which this book offers, and to some confidence in that answer as arising from more than simply impatience, by thinking through what it must now take for us to *earn* our hopes in this arena.

Attenborough rests his, it would seem, on our being the only creatures with foresight: on our having 'an ability, perhaps unique among the living creatures on the planet, to imagine a future and work towards achieving it'. But then it need not come down just to a sardonic as against a cheerful *temperament*, if one points out that we are also the only creatures capable of carrying wishful thinking as far as embedded self-deception, and of deliberately blinding ourselves to a future which we find too ugly to contemplate. Given the conceptual connection between hope and realism, one can get a more objective handle here by pressing the question which of these emphases more realistically reflects the context for hope in our present situation. And that leads on to deeper questioning of the criterion of realism itself.

That is the main line of argument which I pursue here. Very concisely, it runs thus: we have left things so late that honest hope of avoiding climate catastrophe cannot any longer be embraced on anything like Attenborough's terms, but can now only take the form of *hope against hope* – that is, hope deliberately and explicitly embraced in the face of all the stacked-up odds. Such hope can still be realistic, but only with a realism which is not based on appeals to precedent – it must instead reflect a deep-seated faith in the human capacity to create genuinely new possibilities. That faith, however, can only be held onto responsibly, rather than tending to run away into mere utopianism, within a tragic vision of our general condition and current plight. In such a vision, our goods are inexorably conflicted, we cannot tell in advance what values conflict will leave standing, and conflict is in any case bound to be grievous – a matter of severely painful experience as well as of painful choices. And that vision, taken seriously, is also what mandates pursuing urgently necessary transformation at all costs. The demands made by the kind of hope which we can now honestly retain are as tough as that.

I fully expect many people to find this conclusion so unwelcome that they will be tempted to dismiss out of hand the argument leading to it. Naturally, I think that would be mistaken. Of course, the argument – like any argument – may be flawed; but from where we are, it surely has to deserve at least grappling with. Equally, successful revolution in defence of a habitable human future is certainly a very long shot, especially in the kind of society (such as ours in Britain) where it needs first to happen. But I cannot (after much trying) see it as a longer shot than this civilization's reforming itself, at this terribly late stage, by any less-than-revolutionary process. We delude ourselves, surely, if we believe that the speed with which the easy Green agenda has moved from the fringes to the political mainstream over the past forty years tells us anything very much about what will happen now that the really hard agenda has been reached. For all the evidence is that the mainstream just isn't up to embracing it. A recent climate demonstrator's placard in the US urged the official Democratic Party to 'step up or step aside', and that is a neat phrase; but fossil-fuel state politicians, it should now be clear, will never voluntarily do either – that is, if stepping up means acting anything like adequately to address the real dangers. (And this is a criterion by which even the very welcome moves in the right direction by the new Biden administration simply don't qualify.) Those variously invested in the fossil-fuel system will not move more than minimally until the only alternative to action is to be *swept* aside – and many of them not even then.

But that truth, of course, also requires the vanguard itself to step up. From its comfortably accustomed standpoint of informed critique, personal or institutional campaigning or, most adventurously, law-abiding protest (and even getting yourself carefully arrested by a generally well-behaved police force, while of course it calls for courage and determination, is essentially still that), this vanguard must move to becoming capable of *doing* the sweeping aside, and fully ready to do it vigorously.

We need, that is, at last and above all, a Green-political movement adequate to the demands of seeing us through oncoming tragedy – able to interpret inevitable disaster convincingly enough to be widely accepted as a guide, ready to confront painful choices and to cut its losses, remorselessly proactive (since remorse must be tragically forgone, or at any rate suspended), and unflinchingly cold-eyed about its deeper-than-democratic mandate. Accordingly, this book concentrates as fiercely as I can on what I take to be the need and the reasons for that uncompromisingly revolutionary stance. At the worst, therefore, it may have put forward a case to be seriously argued with.

A quick preview

Being realistic about the climate crisis requires us to ask whether, starting from hope as we must, we can keep it both honest and still true to itself, still springing unquenchably from the life which we are called on to defend. Chapter 1 outlines the empirical realism according to which the odds against us appear to be insurmountable. Politics is pre-eminently an arena in which we are liable to ideological bias and groupthink, and correspondingly to self-deception and wishful thinking – with green politics no exception. If, to guard against this tendency, we cultivate a disillusioned realism based on human experience, we must conclude it to be overwhelmingly likely that we are out of time to avoid climate catastrophe. This is demonstrated from consideration of the carbon budget within which we must keep if we are to have any chance of avoiding runaway global heating; of the breakneck rate at which we are still currently overspending that budget; of how precipitously the emissions curve must consequently fall from hereon in; of how inadequate the Paris Agreement has been in this connection; and of how vigorously capitalism and the fossil-fuel state are bound to resist the necessary measures. On all these dimensions, change of the order required is utterly unprecedented, and the utterly unprecedented is the empirically unrealistic.

In Chapter 2, I go on to show how this need not, however, be the final word. The heaviest adverse odds have not always been proof against transformative human action driven by hope 'against hope'. Though history and biography prompt us with this realism of transformation, however, it cannot be grounded empirically. Our ability to defy all the odds cannot be warranted by appeal to experience, which only *warrants* anything through what it shows to be (odds-on) probable. This general point is dramatized especially clearly by our present situation. If we try to believe that we can still save ourselves from climate catastrophe by invoking past occasions (the Blitz, the ending of apartheid ...) when hoping against hope has turned out justified, we find all the analogies failing for the same reason that empirical realism left us disillusioned: the utterly unprecedented nature of our present plight and the sheer scale and rapidity of the change which it calls for. Belief in the power of hope against hope to overcome even that plight must instead be warranted *counter-empirically*, by intuitive recognition of human creativity itself as a fundamental feature of reality.

Knowledge of that creativity goes as deep in us as language, as Chapter 3 tries to make clear. But it has long been overlaid by the resigned empiricist view of human reality which an alienating technologico-utilitarian society fosters. What is involved in recognizing this fundamental creativity as real, and thus in enabling appeal to it to support a kind of realism, can be better understood by exploring the phenomenon of 'hoping against hope'. Whether manifested at the individual or collective level, this involves an acceptance of our first-order hopes as empirically unrealistic, but then, in a deliberate decision to 'hope anyway', weaponizing this acceptance in a way which actually serves to reduce the odds against us. But this process is misrepresented if it is taken to be one of triggering possibilities previously inherent in the given situation. We cannot say that something is already 'on the cards' when, uniquely among the antecedents of natural events, our conscious attention and reaction to the cards has itself to be included among those cards. By making counter-empirical hope practically indispensable, our climate plight thus prompts a new pragmatically-driven approach to the ancient philosophical conundrum of free will.

Chapter 4 urges, controversially, that recovering the creation of possibility in active practice means freeing ourselves from the routine assumption of a *moral* template for our responsibility in this area. Overcoming the collective action dilemma as it relates to climate action cannot be a matter of setting moral obligations against the claims of prudential self-interest, despite the overwhelming prevalence of an ethical framing for climate concern in the literature. This is because the dynamics of the global atmosphere rob all carbon emissions much short of the global total of the causal efficacy required for anything resembling moral constraints to have proper application. That truth has been obscured by the careless metaphor of the 'carbon footprint', the misleading nature of which is here exposed – as far as I am aware, for the first time. The wrong we do by complicity in the fossil-fuel economy and lifestyle must instead be conceived as the betrayal of *life-responsibility*. Self-reinvention as a centre of such responsibility is where our capacity to create new possibilities, and thus to release literally immeasurable forces, must now be exercised.

And such responsibility comes at a price. Unless we are to slide into a magical utopianism (where all hope is indefeasible because failure to attain our objects is just failure to have attained them *yet*), our genuine creativity has to be recognized within the context of an anti-utopian, which means a tragic, view of the human condition. Chapters 5 and 6 make this essential connection. The tragic view understands our values as always liable to deep conflict, and those goods which we realize in action as very often involving the sacrifice of other goods with equal claims on us. This long-established intuitive insight into the inherent limitations of our condition has similarly been occluded by Enlightenment progressivism and its misrepresentation of every difficulty as a problem to be solved. But trust in human creativity needs to be kept honest by understanding its nature and scope within the framing of such a tragic vision. Thus activist hopes invested in unpredictable transformation must retain the *defeasibility* which is the necessary corollary of realism, the *edge* which goes with the real chance of failure. For transformative aspirations, although they can inspire, can also easily betray us if they are not ongoingly adjusted by experience – but counter-empirical hope cannot of its nature be subject to that kind of check. Recognition of the human condition as tragic, it is now argued, must serve as the necessary general constraint on counter-empirical aspiration. That condition is shown to be deeply rooted in our dual nature as rationally evaluative social primates, and arguably indeed in the conditions of life itself.

Chapters 7, 8 and 9 then explore how recovering a tragic world view must make a difference to the way we understand the demands of our immediate situation. The coming years will be hugely and unpredictably turbulent, as people in greater numbers at last begin to register what climate breakdown will involve for themselves and their families, the illegitimacy of shirking governments becomes more and more obvious, and transformative change becomes correspondingly urgent. In the near term only a minority will feel driven to pursue such change, but that minority must come to power *within a couple of decades* if catastrophe is to be averted. How can that be realistic? Can it be compatible with exclusively democratic methods, given what democracy has latterly come to? If not, is that a tragic necessity with which we must live – and how might we do that? What does hope invested in human creativity and kept honest by tragic vision really mean for social and political action in a period for which nothing in history has prepared us?

It is still too early to answer those questions fully with any assurance – the answers which I offer are meant to be exploratory and to provoke debate – but it is by no means too early to ask them. These later chapters therefore try to bring the hope and vision for which the book has argued to bear on key ideas, proposals and campaigns now being canvassed – 'Deep Adaptation', the Green New Deal, Extinction Rebellion – and to consider what form a serious green political movement (which as yet we lack) might take. Thus they may help shape the new kind of revolution to which we must find our way – for nothing less than revolution is what life-hope can now be seen to demand of us.

***

I must add a brief note on methodology. It will probably be already evident from the foregoing preview that I write as a philosopher, although I am certainly not offering what would ordinarily be called a work of philosophy. In particular, this is not 'environmental philosophy' as usually understood: apart from some brief references in passing, there is nothing here about ethical obligations to future generations, other species, wildernesses or whatever. And if it is 'political philosophy', it is so very much in the sense of conceptual work undertaken as a specific intervention in a real and pressing political situation. If I had to classify the book, it would be as a thinking tool for potentially revolutionary climate activists.

The generic kit which philosophers are now widely supposed to bring to the table for this kind of practical purpose is that of conceptual analysis, and the account offered here of our contemporary plight and our room for hope does indeed arise from an exploration of necessary relations between hope, realism and transformative possibility. But at the same time, the role claimed for radical human agency within these conceptual relations calls forth considerations which can't avoid being metaphysical. They can't, that is, avoid seeking directly to identify and engage with the relevant structures of *reality* – quite appropriately, of course, since the aim is to help us think realistically about the climate crisis, and what we take to be realistic thinking must turn finally on what we take to be real.

These 'metaphysical' considerations, however, are not intended to *ground* the rest of the book. Rather, they and the practical concern driving the overall argument should be seen as mutually supportive. The long-standing central 'philosophical problems', I take it, represent the outcropping in different life-arenas – morality, knowledge, personhood, meaning, power – of permanent complexities in our fundamental self-understanding as reflexively conscious, language-using animals. The sense which we make of ourselves in any of these areas both responds to and in turn broadly shapes the operative world-conception of any given historical period and the actions in which it issues. Thus, the picturing of what is *really real* as from an imagined standpoint of objective observation (Thomas Nagel's 'view from nowhere') has long held sway in conjunction with the dominance of physical science and a basically empiricist conception of human nature. This twin dominance, however, with its upshot in the mechanized pursuit of individual material well-being, has now brought us into such deadly ecological danger that it is impossible any longer to avoid confronting it urgently in practice. If our understanding of the real at this epochal crux of action therefore shifts to embrace once more the radical creativity through which human beings can shape possibility, it does so under intense, indeed potentially shattering, life-pressure. But radical creativity is quite unrecognizable unless we take the subjective view – the view from the perspective of the human agent – with ultimate seriousness. And (though at this point I would expect many professionally philosophical colleagues to baulk) one of the fundamental reasons why that subjective view now *has* to be taken fully seriously is that unless we can do so, the unfolding climate emergency will rob us of genuine hope.

The case which I develop through the book aspires to contribute to this kind of shift in understanding. So while it does not mount a metaphysical argument as such, it seeks to keep what I may call the *issue of the real* sufficiently alive that the resonance for this issue of the practical argument which it does mount remains clear – and also *vice versa*. That implies a balancing act which I cannot claim to have performed faultlessly, or even very well, but my attempts to perform it explain some of the apparently awkward shifts in register which the reader may encounter. Overall, the book asks: what might it mean to do the politics of the crucial next twenty years in the spirit of a tragically informed realism? Some of the answers which I offer will certainly be found as uncomfortable as the question is exigent. The *full* force of those answers requires, I believe, that we appreciate something as fundamental as the nature of the real to be always also at stake.

★★★

Introductions to books about hope and the climate crisis standardly end in upbeat vein: we can crack this, if only … (followed by the author's particular hopeful pitch). But prefacing a critique of the possibilities of hope in that way would evidently beg the question. Instead, I will invoke the book's dedication: jointly to my mother who died while it was being written, and to my first grandchild whose birth spurred me into writing it. That life-continuity can perhaps stand in hope's place until hope is vindicated – or, while one at least attempts its vindication.

Hallbankgate, Cumbria
May 2021

1

The Demands of Realism

The condition of realism, as I stated it in the Introduction, was that hope must address itself to a real chance of the hoped-for event's coming to pass. But what counts as a real chance? Here, already, we need to embark on some conceptual clarification.

The odds are apparently about a million to one, for instance, against your being struck by lightning in any given year. Is there, nevertheless, a real chance that this will happen to you? There is a sense in which that assertion of the odds itself constitutes an affirmative answer: yes, indeed, there *really is* such a chance, although it is a very, very slim one because misfortunes of this nature occur extremely rarely – only about one time in a million when you are out and about, to be more precise. But then, in that sense, the condition of realism would be met by any case where what one was hoping for wasn't something literally impossible, such as travelling back in time, however overwhelming one judged the odds against its happening to be – and that is plainly not what we intend when we recognize that hope has to be realistic to do its proper work.

Instead, we mean 'realistic' here in the sense in which it would be *un*realistic to frame one's New Year resolutions, holiday arrangements and so forth on the basis that one might on any day in the coming twelve months be taken out by lightning – just as it would be still less realistic to erect such life-plans on the firm assumption that one will this year win millions on the Lottery. The chances are such in either case that it would be wildly impractical to spend any time worrying about or joyously anticipating the events in question, or even (except perhaps idly and momentarily) contemplating them at all. The condition of realism in relation to both our fears and our hopes, that is, carries an essential reference to the common-sense distinction between attention to the way the world objectively is, and fearful or *wishful thinking*.

Wishful thinking, which this opposition distinguishes specifically from hope, we could define for the moment as the process of letting one's desires play a larger role in generating one's expectations about what will happen than the facts relevant to what is likely to happen actually warrant. That is perhaps a possibility for any creature which has both cognitions and desires and is at least minimally aware of both. Any such creature seems susceptible in principle to finding itself misled by its own wishfulness. But human beings are unique in having at their disposal very sophisticated powers of misrepresenting the world – of seeing what they want to see and believing what it comforts them to believe. The experimental chimp can expect a banana and get an electric shock, the dog can bounce around in anticipation of a walk and find himself left behind when his owner is merely going shopping, but it takes human intelligence and conceptual resource to weave a dreamland of protected quasi-beliefs and gratifyingly false assumptions which can be inhabited for weeks, months or even years together. And by the same token, it would seem to be principally the human ability to recognize the patterns, rules and norms (including laws of nature) derived from remembered experience, which provides us with any reliable benchmark for our attempts to resist losing touch in this way with the real world.

Avoiding illusion

We misrepresent things wishfully to ourselves across many diverse fields of activity and for many reasons. Standard cases are to be found in our personal relations, where we regularly promote muddle and grief by assigning fantasy roles to real others, and in our attitudes to our own achievements, professional and other, where the fantasy role is typically filled by oneself. (Familiarly, one of the toughest things about growing older is that it becomes ever harder to find examples of successful people who didn't amount to much at your age.) But because it deals in *power* (ours over others and theirs over us), politics is a pre-eminent terrain of systematic self-delusion – of vision through deliberately assumed ideological spectacles, of beliefs and drives channelled by class or clan interests and of the group-conditioned reflex generally. George Orwell's classic insight remains sharply relevant here. In an article written just after the Second World War, when he was already exploring the ideas that were to inform the chilling satirical account of 'doublethink' in *Nineteen Eighty-Four*, he observed drily that

> In private life most people are fairly realistic. When one is making out one's weekly budget, two and two invariably make four. Politics, on the other hand, is a sort of sub-atomic or non-Euclidean world where it is quite easy for the part to be greater than the whole or for two objects to be in the same place simultaneously.

Politics, that is, characteristically constitutes an arena where one is at risk of understanding only what one wants or allows oneself to understand, even where this conflicts with what one already, in some other part of one's mind, does tacitly see and understand.

As a direct consequence of this, history insofar as it is a record of political activity exemplifies a very wide variety of unrealisms, and is equally rich in cases of people only finding out that they were being unrealistic *afterwards*, when action has collided painfully with the resistant reality which they had misconstrued or downplayed. And of course it also displays abundantly the irony of our never knowing beforehand whether or not that is actually going to be our situation. Thus, to take a couple of instances more or less at random, while it turned out to have been realistic of Hitler to think that he could come to power in Weimar Germany, despite his widespread dismissal during much of the 1920s as a political no-hoper, it proved wildly unrealistic of him to suppose that he could knock out Soviet Russia at a single blow in 1941, even though no army in Europe had been able to resist the Wehrmacht for years before. Again, it was in fact realistic of Tony Blair to trust to his instincts and push for the abandonment of Labour's time-hallowed Clause Four commitment to public ownership almost immediately after taking over as leader, but career-wreckingly unrealistic of him to imagine that he could inspire public support for the invasion of Iraq solid enough to tolerate anything except the swiftest and least costly victory and exit. In both these pairs of examples and in a thousand others like them, calculations beforehand easily deceive, and the realistic option, the one that would have reflected how the balance of relevant political and military forces actually stood, is revealed only when the chosen alternative has led to confusion, chaos or disaster.

In political and public affairs, however, we much more often than not have to act well before the relevant reality has fully declared itself. Thus our constant liability to falsify our situation for Orwell's kind of reason tends to place us even more than usually at the mercy of events in these spheres. As a defensive response, there have therefore naturally evolved some fairly straightforward rules of thumb (if only we could stick to them) for aligning one's picture of how things are with how they actually are, *before* acting on it. These rules of thumb familiarly include paying less attention to what people *say* (about what they believe, intend and so on) than to what they have actually done, and particularly to what they have been in the regular habit of doing; assuming that, generally speaking, more immediate and self-interested concerns will win out over remoter and more altruistic ones; recognizing that, of two conflicting 'expert' opinions, the reassuring one will always be much more readily believed; and trying hard not to underestimate people's capacities for self-deception about motives, loyalties, values and goals. This response needn't go so far as the *realpolitik* which takes it for granted that human beings are always inherently egoistic and self-interested, and that states naturally replicate those motives in their dealings with one another. But the general strategy is nevertheless to step back as far as possible from whatever is liable to be infected by self-deceiving or self-gratifying misrepresentation, in order to identify patterns of activity which have a chance of inhering objectively in the kinds of situation which again and again come up. It acknowledges that, while in human affairs we are never operating under anything like laboratory conditions with surrounding circumstances held artificially constant, nevertheless an objectively reliable way to keep ourselves in contact with how matters actually stand is to ask what broad patterns we can perceive in our experience, and base our future actions on projecting those patterns forward. This whole empirically-grounded interpretive mindset, or apparatus of pragmatic suspicion, we might well call the realism of *disillusion*. It reflects long-embedded human experience of having been regularly disillusioned whenever ideology or blinkered groupthink has first obscured and then collided with reality; and it seeks deliberately to puncture in advance any further such illusions under which we might be labouring going forward.

Now evidently our climate and environmental future is an intensely political matter. It is also something which we daren't misrepresent to ourselves if we can possibly avoid doing so, especially at this now very late stage – the stakes (the habitability of the planet) being just too high. So there is a premium in this area, perhaps above all others, on bringing a determinedly disillusioned eye to the scrutiny of our prospects – that is, on submitting our hopes to the test of a firmly empirical realism, in order to distinguish whatever may be robust in them from whatever is a product of wishful thinking.

What, then, does such a scrutiny reveal?

The vicious syllogism

I think it now plainly tells us that if we *had* been going to prevent destructive climate change, we would have put in place genuine constraints on our emissions-generating behaviour worldwide quite soon after this first became a live issue, instead of dragging our collective feet from the 1980s through Rio 1992 to Katowice 2018 and beyond, until it was much too late; but we didn't; and so, we're not going to prevent it. This little argument, which in *After Sustainability* I called 'the vicious syllogism', is valid (that is to say, its conclusion is guaranteed if its premises are true) and its minor premise, at any rate, clearly *is* true. To the disillusioned eye, the vision informed by carefully rejecting the temptations of illusion, we palpably *didn't* do what was needed by way of putting serious constraints in place when we had the chance. We learnt to talk the talk over this period, but only ever to walk as much of the walk as would enable us to go on talking – as the sad succession of last-chance saloons along the way (Copenhagen, Cancun … Paris) makes plain enough. The form standardly taken by wishful thinking over this period, as

I suggested in the Introduction, has been that of 'sustainable development'. This approach has not only failed to deliver change of anything like the order required since it started to figure in mainstream policymaking: it was evidently always going to fail. It looks hard-headedly pragmatic at first blush: having quantified future needs for ecological resources to sustain welfare levels at least equal to our own, we work back, again quantifiably, to what we must do or refrain from doing now to ensure that those future needs can be met. But this whole purported constraint from the future depends on our believing what we want to believe, because under pressure it is as full of escape clauses as a dodgy insurance contract. Its apparent pragmatism turns on gross exaggeration of our powers to predict and control, and does so in the service of supposed stewardship obligations to future generations which are no more than pseudo-obligations, since the only people interpreting them and monitoring how far we are meeting them are ourselves – that is, present people who are obviously interested parties. Its basic purpose, and the real reason why environmental concern got mainstreamed on this model, has been to ensure that we have been able to continue wishfully shunting the last-chance saloon-car further on down the track without ever banging up against any buffers *within the model* – although, of course, the real-world buffers are rapidly looming up.

Meanwhile, the hypothetical major premise of the vicious syllogism ("*If* we had been going to …") depends on the assumption that we are now out of time for the required changes, and thus looks as well grounded in scientific evidence and hard-headed economic, sociological and political observation as any empirically based counterfactual well could be. Perhaps the most effective way to demonstrate this point is to invoke the concept of the global carbon budget. What warms the atmosphere by trapping solar heat (the 'greenhouse effect') is basically the accumulation of carbon dioxide resulting from increasingly carbon-intensive human activity since the start of the Industrial Revolution. Focusing on the cumulative nature of these emissions, we can easily think of the difference between the amount of human-generated CO_2 already aloft, and the amounts which will, as they are reached, produce various degrees of warming above the Earth's pre-industrial state: 1.5°C, 2°C and so on upwards (there being a reasonably linear relationship between the atmosphere's CO_2 burden and global surface temperature). These differences are then our still-available carbon budgets, the amounts of extra emitted CO_2 to which we must limit ourselves if we want to retain a calculable chance of keeping atmospheric temperature rise to each of those identified levels in turn. The current scientific consensus at the time of writing (September 2020) is that we must live within a budget of *at most* around 495 billion more tons of emitted CO_2 (some expert estimates suggest a considerably lower figure) in order to have an evens chance of keeping warming below 1.5°C – the level which has latterly been taken by the climate policy community, led by the Intergovernmental Panel on

Climate Change (IPCC), as just about manageable for the ongoing viability of global civilization in something recognizably like its present form.

Four hundred and ninety-five *billion* tons of CO_2 sounds a lot – enough, perhaps, to give us time for adaptation, for developing the cleaner technologies, carbon capture and storage systems and widespread substitution of renewables for fossil fuels which we will need either to keep within our carbon budget for the foreseeable future, or to move us into a zero-carbon regime when the budget runs out. Visit the Carbon Countdown website maintained by *The Guardian* newspaper, however, and that impression of global elbow-room evaporates. This web page displays figures, changing in real time, for the total remaining carbon budget for a 66 per cent chance of 2°C; that is, the target level to which the Paris accords supposedly committed everyone, 1.5°C being still aspirational at that point although it has since been recognized as the level at which the commitment should have been set. The carbon budget for the higher figure is of course somewhat larger, at around 647 billion tons, but there still only remain some sixteen and a half years before at current emissions rates we overspend it. That is perhaps startlingly less than one might have expected, but the *really* shocking figure is the one for ongoing emissions. Confronting you at each visit to the site is a countdown clock showing the worldwide total in tons of CO_2 equivalent which has been emitted since you clicked in. As the numbers on this clock flicker past, so swiftly that you can only keep track of the mounting thousands, the scale of the challenge facing humanity emerges literally before your eyes. At each of several visits during the period of writing, this counter was registering atmospheric carbon being added worldwide at the rate of some 77,000 tons per minute, or about 1,280 tons – a weight of carbon roughly equal to that of 103 new-style Routemaster buses – being pumped up into the skies *every second*. (That's some four thousand buses-worth since you started reading this paragraph, unless you can speed-read.) And when you remind yourself that atmospheric carbon dioxide is a *gas*, and therefore how much of it would be required to tot up to just a single ton in weight, and how far (for instance) you personally would have to drive in order to emit even that much, you can't help being overwhelmed by a sense of the gigantic scale, utter relentlessness and frenzied intensity at which human activity is spewing out this climate-destabilizing pollutant around the globe.

And of course by the same token, these eye-glazing magnitudes emphasize the scale and rapidity of the changes in all this activity which would be required if we are to have any prospect of hitting even the higher (Paris) target. An effective graphic representation of what would be involved is at Figure 1, which pictures what would be required to give us a slightly better than 66 per cent, but still not especially confidence-inspiring, chance of keeping within 2°C. (It is worth asking yourself whether you would get on a plane with this chance of not crashing *en route*.) Even on the very

Figure 1: The carbon cliff-edge: global carbon dioxide cuts required for a 75 per cent chance of not exceeding 2°C, with a peak today

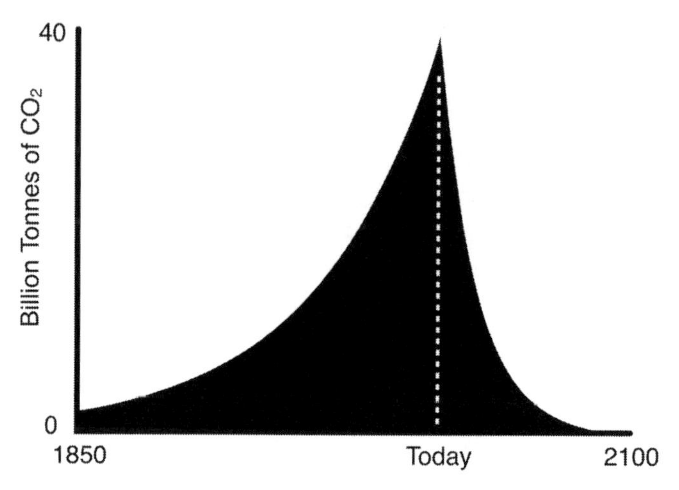

Source: Image from *The Burning Question: We Can't Burn Half the World's Oil, Coal and Gas: So How Do We Quit?* by Mike Berners-Lee and Duncan Clark, foreword by Bill McKibben. Reprinted with permission from Profile Books and Greystone Books.

optimistic assumption that emissions have now peaked, the precipitously plunging curve from that peak gives to anyone who has ever ridden a fairground roller coaster a very vivid impression of what the necessary changes would feel like. Professor Kevin Anderson of the University of Manchester and the UK's Tyndall Centre for Climate Change Research has argued that in pursuing these changes we should be looking at global energy-related CO_2 emissions decreasing by 10–20 per cent per year over the next decades: 'Flying, driving, heating our homes, using our appliances, basically everything we do would need to be zero carbon.' And plainly, to achieve such cuts would mean economic and social upheaval on a scale on which the 'austerity' imposed in the UK after 2010, for instance, simply wouldn't register.

Nor would this be just (just!) a matter of people worldwide altering their directly energy-using behaviours. As I noted in the Introduction, a crucial driving force for dangerously high levels of global CO_2 emissions at the present time is the heavy reliance on coal-based electricity generation in the rapidly growing economies of China and India (among others countries, but these are the principal ones). Simplifying boldly, we can identify two main motive forces for this growth. First, there is the desire of the peoples of these countries to attain something much closer to Western material standards of living, as exhibited to them by a globalized media, than traditionally they have enjoyed. Second, the deindustrialization of many Western economies, such as the British, in recent decades has exported much production – and therefore, much of the energy-demands of production – of the material goods supporting their favoured lifestyles to these countries where labour costs are cheaper. (This process, which essentially transfers Western responsibility for carbon emissions eastwards, is the reason why it is unjustifiable, even if it were a practical policy, to *blame* China – in particular – for the intractability of the climate crisis.) There would seem from our entire track record to date to be no credible scenario under which all this indirect energy demand is shifted to renewables within anything like the necessary time frame and given current renewables technology. Hence, there would equally seem to be no chance whatever of even beginning to reverse these trends while the West remains firmly attached to a way of life heavily dependent on material goods (and their cross-global transit) as well as in itself very energy-intensive. Hence, in turn, the vitally necessary changes represented by the plunging curve must also include widespread abandonment of the consumerist lifestyles and expectations which are currently interlocked with global capitalism, and correspondingly a very significant localization of power to communities and away from global elites with vested interests in the centralized fossil-fuel state. That means a decisive shift away from neo-liberal capitalism as the default socio-economic mode in the North and West of the world – even if we factor in the controlled roll-out of negative emissions technologies (themselves still very speculative) for reabsorbing already-emitted carbon and do not rely on these as justification for resiling from any of the other requirements, but essentially use them as 'belt and braces'. Nor would there be any real prospect that even this epochal recension of priorities would be effective without an accompanying and very significant change in Northern and Western attitudes to overseas aid aimed at mitigating at least some of the impacts of global heating already in the system on the areas of the world most exposed to those consequences, as well as at building renewable capacity and local economic resilience in those areas.

Attempting to envisage shifts of this order, in all their geopolitical, economic, organizational, cultural and lifestyle dimensions, makes it irresistibly clear not just how drastic are the changes in question, but also that change at such breakneck speed is without the faintest shadow of a forerunner in human affairs. As Anderson soberly puts it, 'we simply have no precedent for transforming our economies in line with our commitments to avoid dangerous … climate change'. *Nothing* in any human experience to date suggests that change of that order in that kind of timescale could be practically possible.

A bleak prospect

So the claim that we can't now turn things around and avoid a succession of climate-driven disasters within the window of opportunity which we have left ourselves appears to be decisively vindicated by these empirically realistic considerations. This is an ugly conclusion, but one which recent thinkers about the issue have increasingly been reaching. As well as my own work, a wide diversity of authors (Anderson, as mentioned, together with Jem Bendell, Clive Hamilton, Dale Jamieson, Tim Mulgan, Rupert Read and David Wallace-Wells) now make the working assumption that climate change, somewhere between seriously disruptive and catastrophic, is no longer something we must find ways of avoiding, but something we are going to have to live with. Parallel to this recognition is the rise of the 'Anthropocene' trope with its defining acceptance that human beings have decisively altered the atmosphere and set in motion a mass extinction as drastic as any produced by Earth-system changes over geological time.

Ah, but – someone might protest – what about the Paris Agreement? Hasn't that been the game-changer which gives us a better chance? As Brian Heatley and Rupert Read point out, however, in a paper from the think tank Green House, achievement of the Paris goals depends entirely on the self-monitored meeting by each signatory country of its self-set emissions reduction targets. The apparent plausibility of these targets is largely an artefact of setting a modest-seeming annual reduction commitment against a target date which still feels comfortably distant – and we need only recall the racing carbon-counter to have the gravest doubts about *that*. But we shouldn't in any case need that reminder to make us realistically cynical about such a process, when entered upon by countries still uniformly committed to economic growth and rising material living standards, and therefore resembling nothing so much as a bunch of alcoholics left to agree and police their own safe drinking levels. Subsequent developments up to and including the 2018 Conference of the Parties at Katowice (focused on fiddling the rule book while Rome burns) provide no grounds at all for modifying that comparison.

Moreover, even if such a process inspired any confidence (and as climate scientists had in fact pointed out well beforehand), the whole regime based on the 2°C target is still called into question by highly likely feedback effects of which it takes no account, such as release of methane from thawing permafrost or increased heat absorption by the ocean as the Arctic ice melts. So even though the rogue Trump administration didn't manage to wreck the whole Paris-initiated process, the clear implication of all this to a disillusioned eye is that global temperature rise will be at the very least 3–4°C over pre-industrial levels by 2100, and more likely 4–5°C – taking us into a world which has been described by those in a position to know as 'incompatible with any reasonable characterisation of an organised, equitable and civilised global community … [and] devastating for many if not the majority of ecosystems'.

It is no argument at all against an empirical realism this bleak that it goes unacknowledged, and still less acted on, in the official policy world.

The truths which it would force us to acknowledge are very hard, and modern telly-democratic electoral politics, with its crass simplifications, brief attention spans and routine substitution of slogans for thought, has not characteristically been an arena in which hard truths can be faced and then fed into policymaking. That we are already in for what a former UK government Chief Scientific Adviser, Sir John Beddington, has described as a 'perfect storm' of food, water and energy shortages, entailing famine, disease and homelessness on an epic scale, with associated worldwide migratory pressures and resource wars, is not going to form the starting point of any election manifesto under present political conditions – and this includes the manifestos of Green parties which have undertaken to operate under those conditions. It is far easier for the political and policy community to stay in denial with the paradigm of 'sustainable development', and its climate-change corollary of managed incremental (or would-be incremental) emissions constraints. I have commented already on what this paradigm does for us by way of giving comfort to wishful thinking. What it certainly doesn't do is mandate messy and dangerous challenges to global corporate power, the world trade regulations which seek to lock that power into place, nor the governing elites who profit so grotesquely by these arrangements. Hence, in large measure, its continuing official hegemony. Appropriate action in the public realm would, as I have shown, need to be driven by policies for the decisive ending and indeed reversal of material 'progress' and for the best available mitigation of a now gravely alarming climate and ecological future. No such policies are imaginable from any plausible government of any fossil-fuel state as currently constituted. So it looks as if what is already firmly on course to be disastrous won't be addressed, or even recognized, in time to prevent its escalating into catastrophe.

How might this escalation progress through the lifetime of someone born around now? We could divide the process roughly into three phases, the first of these characterized by widespread and various forms of denying the situation's seriousness, until the reality of disastrous impacts becomes unignorable; then a desperate resort to heroic technology and authoritarian political management: the failure of which will deliver us into an increasingly Hobbesian world of the kind imagined in John Lanchester's recent novel *The Wall*, where getting one's survival strategy in first is the only thing that really matters to nations and increasingly to individuals. This division is illustrative and all the hazards of forecasting must be remembered. That said, it is empirically realistic to suppose that in the first of these phases, the world will by and large keep soldiering on within the Paris framework, or patched-up successors based on equally flawed premises, because facing up to the inadequacy of that approach would simply demand too huge a political and lifestyle shift. As this strategy fails, however, and the 'perfect storm' which Beddington anticipated from the impacts of continually rising emissions starts to become unignorable (he dated this development at somewhere around 2030), responses in the second phase will include a combination of command-and-control energy substitution (too little too late), border controls against forced migration, resource wars for the remaining fossil fuels and predictably mismanaged geo-engineering interventions, while the Paris-plus regime finally unravels. This trajectory will not be simply business as usual because it will take place within a context of at least would-be mitigation, but it will still be mitigation designed, on the 'sustainability' model, for pretending that we are doing far more than actually we are, and so for achieving much less than is called for. It will take us maybe to a 3–4°C world rather than the 4–5°C which would have resulted from business as usual.

By around mid-century, however, massive feedback effects from our failure to control global heating to 2°C up to that point will have started to become evident. These will arise specifically from the release of methane from thawing permafrost, the effects of rainforest dieback and ocean warming in reducing carbon dioxide absorption and significantly rising sea levels from melting polar ice. Their consequences will entail sea-level rise of at least 3–4 feet with many low-lying coastal areas, including in Europe and the US, becoming uninhabitable; many more extreme weather events such as major storms and widespread flooding; high temperatures over land making outdoor work difficult or impossible in some months of the year, with dustbowl conditions in many areas currently heavily populated or intensively farmed, and resulting food and water insecurity leading to severe famine and drought in high-population areas within the global South and a range of direct and indirect health impacts elsewhere.

While the broad parameters of these effects have been explored in scientific modelling which has already been done, they are individually unpredictable and could well be more severe than this suggests. (The general trend is for improved modelling to show up previous figures as underestimates.) Even less predictable are the ways in which they will interact with each other and with the changes that will already have been set in train in the first half of the century by the break-up of the United Nations Framework Convention on Climate Change (UNFCCC). Plausibly to be anticipated on the basis of the historical record are wars over resources and territorial defence which will massively destabilize the international balance of power and global security – and it will be a significant stroke of luck if some of these wars don't involve a nuclear dimension.

Disillusioned hope … ?

That, if we are looking to be empirically realistic, is the future for my grandchildren – and of course for the rest of their generation across the world and those generations which will follow. It is a bleak prospect indeed. Is it also *hopeless*?

With this question we are prompted to explore further into the idea of hope than the brief foray in the Introduction took us. I said there that hope for the human future is not optional, and linked that claim to its arising out of the energy of life itself. But there are of course a variety of possible futures, including the one just sketched, through which life in its human manifestation might persist. Is there anything in the nature of hope to demand that we invest it in something more ambitious than whatever a disillusioned empiricism suggests is realistic? My whole argument in this book is going to turn on an affirmative answer here.

Hope, as I suggested in those introductory remarks, is the form taken for a creature which can think about its future by life's instinctive always-ongoingness. Life's commitment to itself takes, in creatures conscious of having a future into which they are always moving, the form of a forward-facing generic trust. That trust exhibits itself to us as hopefulness in respect of the particular projects in which we invest our life-energies. And because having that kind of consciousness of self over time means being aware of one's exposure to contingency, the condition of realism arises in connection with it: we cannot just trust blindly to life, as creatures without conscious foresight do, but intentionally direct our living forwards through hope invested in our various projects and purposes. Where we invest it is then guided by what we are aware of being able realistically to hope for, and that is how we steer ourselves through a world of contingency: where hope in ordinary cases evidently fails the condition of realism, we withdraw it and transfer it elsewhere as part of the very process of conscious life moving forward.

But we cannot withdraw hope in that way from the overall project of life, which expresses itself precisely as generic hopefulness seeking out plausible projects. Thus we have no option but to hope that we can overcome threats to the viability of the human life which manifests itself in us *as* hope. That is the fundamental sense in which we must take the claim that hope is not optional in face of the climate emergency – life itself in us will not let us just give up on any prospect of a human future and enjoy what is left to us in the present while we can.

It is to this sense, I think, that the basic notion of *sustainability* must be related, if we are not to misconceive it. That concept tends to be interpreted economistically as the intergenerational transmission of equal resource-quanta. But its compelling force as a goal for action comes from much deeper down – from recognition that it expresses the energy of continuity at the core of life. The hope that what we do can be sustainable at bottom expresses an intuitive awareness that a crucial part of the point of life is just to iterate itself onwards. The underlying impulse of sustainability, in other words, *is* life-hope.

But then, to press again in a different form the question from the start of this section: how much does sustainability, so understood, demand? Are the very reduced prospects which seem to be all that empirical realism now leaves us, hopeless in that sense?

On these grounds, surely not. Life and hope seem so connected that where life remains at all, no prospect can be completely hopeless, and we are certainly not here confronting the final end, either of human or – despite the forecast sixth mass extinction – of all other life either. Moreover, it seems possible for at least some of us to go on hoping that anyone about whom one actually cares will belong to a lucky minority. Those will be people living in currently temperate zones which will become warmer but still for a time remain habitable. They will need to live outside urban areas where life-support systems will collapse, and beyond the suburban wastelands in which people will find themselves stranded as transport infrastructure unravels. They will need somewhere with some access to land which can be made productive and on which well-integrated, intelligently led and practicably defensible local communities can be reconstituted. (Continuingly habitable areas will in this scenario depend on self-sufficiency not only in food and other necessary goods, but also in effective weaponry.) And these lucky ones may be able to insulate themselves from the wider global effects of climate destabilization for some considerable time. Thinking parochially (as in respect of such a future one has little option but to do), this is all fairly readily imaginable for at least parts of Britain, which is not only temperate and even yet incompletely suburbanized, but which starts the process of insulating itself from wider chaos with the great advantage of being already an island. (See, again, Lanchester's story *The Wall.*) This prospect is grim, but it is plainly distinct from the classic post-nuclear-holocaust situation where, in Khrushchev's chilling prediction, 'the living would envy the dead'. We should surely be foolish to claim that life under these conditions – at the favoured end of the feasible – could never be worth living in default of anything better.

It would nevertheless clearly involve hugely diminished human possibilities, even for the lucky minority. Our recognition of and reaction to this reflects our strong tendency, already noted, to think of what sustainability requires in moral terms, and specifically in terms of *justice* – as placing us under an obligation to leave future people no worse off, at least in essentials, than we currently find ourselves. (Hence, of course, arises the dominant picture of this requirement as to be operationalized economistically.) There seems to be an obligation of this kind on us in the present to direct our hopes towards a much more generous future. Thus to envisage a world in which there is any human flourishing at all would involve hoping for one where nobody, wherever or *when*ever born, starves or suffers systematic deprivation merely by that chance of birth; a world where everyone enjoys, as of right, a normal human lifespan in reasonable health, is able to gain respect and obtain reasonably fulfilling work and has the chance of finding a life-partner and raising a family. Plainly, from the perspective of disillusioned empirical realism, this is not what a destabilizing climate and a degrading biosphere are going to permit, as the twenty-first century advances towards the twenty-second, to any but a lucky minority, who would therefore be most unjustly favoured by geography and history. Moreover, if the only hope which such realism leaves us is that these lucky ones might continue (for a while) to enjoy some limited material well-being, that might seem to be giving up on human flourishing as such. For it is in the nature of human beings to flourish, if they do, as self-aware moral agents, and in the kind of moral regime envisaged, where those remaining in the habitable interstices of an irreparably damaged world would depend for their survival on the abandonment of others to destitution, famine and war, any prospect of human flourishing might seem hopeless.

Simply as far as that goes, however, it could well be responded that such a view allows the best to become the enemy of the sensibly pragmatic. It is surely better, even *morally* better in a longer view, that some people continue to live tolerable lives, albeit under hugely adverse global conditions, than that none do – if only because that preserves the outside chance of gradually turning things around again for greater numbers, and at least maintains some continuity of civilization in the meanwhile, albeit a form of civilization for which no universalist morality remains possible. If the global economy collapses quickly enough under pressure from accumulating climate disasters, the warming to which we are already committed just might not trigger positive feedback drastic enough to make the whole planet permanently incapable of supporting organized human life, so that a narrow pathway through to a broken but still partially habitable world may remain open. And while a just world is no doubt ideally best, one in which some people retain enough elbow room for keeping alive the aspiration to justice along with other characteristic features of the human project seems as if it must be better than nothing. Hope for the better still makes sense even when it must be entertained in full recognition that the best is beyond our reach.

And as regards our present moral obligation, as decent people, to hope for something more generous, it has to be admitted that morality in this context now cuts, very sharply, both ways. For as well as a requirement to pursue and affirm justice, we are also under a moral requirement to be honest. Honesty, as I have been pointing out all through this chapter, demands that we do not protect our hopes for a just future world from failure to meet the condition of realism by indulging ourselves in any of the multiple forms of activist denial which are so readily available – extending all the way from the broken-backed sustainable development paradigm at the policy level, to the manifold displacement activities with which environmentalism prompts us as individuals. I have written elsewhere about denial of this kind. Unlike the still too-common denialism driven by financial and other interests vested in not admitting anthropogenic climate change to be serious or even real – driven, that is, by greed where not by mere ignorance – it demands respect: those who resort to it do so to make space for honourable hopes. But for all that, its general form is the attitude classically expressed when a recent survey question asked respondents whether they thought it was too late to stop drastic climate change: 'No, of course it's not too late, because if you think it's too late, then where's the drive to act immediately?' If we are being empirically realistic, this is just the chocolate-cake logic of wishful thinking ("Of course it's not too fattening, because if I thought it was too fattening, I might have to not eat it"). The danger of deploying that logic is not so much that you risk putting on weight, as that you do so through forsaking contact with reality. And then in the climate case, the hope for which you might thus have made room turns into an insidious kind of self-indulgence – it has come to be more about what you can bear to accept and confront, than about its ostensible objects. The bottom line is that the moral requirement of being honest here must trump that of pursuing justice, since there is at least some chance of meeting it.

So realism might seem bound to confine our hopes to these very disillusioned prospects. But this is before we reckon further with the inner dynamic of life-hope itself – hope as it both expresses life-energy and engages with the conditions of life-continuity. Such hope *of its very nature* demands to be directed at sustainability in its much fuller sense: human flourishing within a rich and diverse biosphere indefinitely into the future.

... or life-hope?

This can be seen most clearly by foregrounding the impersonal dimension beyond the personal concern when, say, I worry about the future awaiting my infant grandson. What do I hope for him? Well, first of all the basics – conditions in which he can live healthily and in physical safety, and so on. I am also concerned that he should be able to find meaning and purpose, encounter beauty and experience wonder. Being aware that humans cannot flourish in any of these ways without a functioning biosphere surrounding and supporting them, I also hope that this will continue to be the case while he lives. But the key point is that I don't hope all this for him just as some particular individual about whose welfare I happen to be exercised, but as someone for whom I have a *grandfatherly* concern – and that is, a concern arising directly out of the natural ongoingness of life. That makes a crucial difference. For if what I hope for on his behalf arises thus, it would be completely self-defeating not to hope for it as something which I must also at the same time hope that he will, in his turn, be in a position to hope for in respect of his own children and grandchildren, and then they for theirs.

I take over this thought gratefully from my fellow philosopher Rupert Read, though I give it a form slightly different from that in which he has expressed it. His version of the argument is that if we care for our children, we must care equally for whatever they will care for, which will mean caring in just the same way for our grandchildren, for our great-grandchildren and so onwards. Our care for the *n*th generation will always therefore be just as intense as that which we feel for our immediate descendants, and this will motivate us to avoid harming *any* future generation as we should flinch from harming our own children. This avoidance will then take the form of our adopting present practices premised on indefinite biospheric integrity. That formulation in terms of *care*, I have to confess, seems to me implausible. Care is surely the kind of feeling which quite naturally attenuates with either physical or temporal distance, and I am perfectly happy to admit to not caring a hoot for my thirtieth-century descendants, if any. But my present grandfatherly concern for my grandson's flourishing does I think commit me in the way I have suggested to desiring (now) the indefinitely continuing sustainability, as we should try to retrieve the term from economistic misuse by calling it, of humans-in-the-biosphere. This plainly cannot just be a matter of his managing to get through and out, nor of his own grandchildren doing so, before a degrading biosphere becomes terminally damaged. Rather, I have to hope for the continuance of a healthy relation between human beings and the biosphere as the basis of life's essential ongoing *intergenerationality*, its iterating itself onwards as far as we can foresee, and further yet. And my hope for this is not born from any pseudo-obligation owed to distantly future people, with all the baggage of conceptual difficulties which that idea drags in its train, but from a genuine duty owed to my children and grandchildren, to whom I am actually answerable in the way of intergenerationality. To put this in something like Read's terms: if you really love your kids, you owe them your best efforts to retain an *indefinitely* habitable planet.

Now a wide variety of possible ways of life, economic arrangements and distributions of power, resources and opportunity could be consonant with that onward iteration of habitability (so that concerns for justice, *just as such*, are beside this particular point). But what certainly could not be so consonant is the sort of human-driven further degradation of the living world which a disillusioned empirical realism, as outlined previously, now gives us overwhelming warrant to expect. Accepting such a future as all we have left to hope for would involve a deep betrayal of the life-continuity which life's inherent intergenerationality itself charges us to foster.

'Grandfatherly concern', in other words, is not a concern just for the appealing little individual who happens to be my grandson, but for how in and through him life moves forward from ancestor to descendant becoming in turn ancestor, and so onwards. It is an impersonal concern which demands robustly stable continuity in the biosphere within which human life has always been so intimately embedded. (It is not just for technical reasons that we could not export ourselves to carry on human life in a geodesic dome on an alien planet.) More, it demands such a biosphere non-instrumentally. Just as hope for my grandson's flourishing cannot stop with him, but invests itself in the indefinite continuance of a human heritage in which he and I are linked, as an expression of impersonal life-impulse flowing through us, so life-hope cannot consort with an attitude to the biosphere which regards it as purely instrumental to human flourishing – as having done its job, as it were, provided humanly habitable enclaves remain. Hope invested in human heritage has of its nature to be more non-anthropocentrically generous than that: its impersonality in relation to particular human individuals is denatured unless it carries over into non-specificity in the literal sense, a non-species-centred hope for a world of resilient and robustly generative (which means, prolifically diverse) life-at-large. We can't hope thus for an ongoing human heritage subserved by a destabilized and precarious biosphere which manages to supply us with just enough patchy support for viability. Sustainability, as in the nature of hope and life we have to hope for it, has to mean indefinite human flourishing within an *independently* rich and diverse living world. The hope that my grandson's grandson, and his in turn, will be able to hear the curlew's call from the moorland, and the hope that the curlew will go on calling in springtime whether or not anyone hears it, are both expressions, in the uniquely human mode of hope, of the same undifferentiated and indefeasible life-energy which vibrates in us, in the curlew and in all life.

But this brings us to the crux which it has been the point of this whole chapter to make explicit. We have seen that investing hope in the human future is not optional, not something which we can ever really abandon while we remain alive, and also that it must consciously meet the condition of realism if it is to maintain itself as hope rather than collapsing into wishful thinking. Yet we have seen also that the internal logic of life-hope requires us to envisage for that future a kind of sustainable world, the transition to which it is, if we are honest, no longer realistic to regard as achievable, *as long as we understand realism as shaped by what is empirically plausible.*

This bind is uninhabitable – not only intellectually, but also vitally. The only way out of it is to appeal to a different way of understanding realism. The next chapter begins that task.

2

Transformation?

The realism of empirical plausibility represents, as I said in the Introduction, our taking the parameters for what is *count* as thinking realistically from a certain underlying conception of, or set of unquestioned assumptions about, what is fundamentally real. Central to this at least tacitly operative metaphysic is the belief that in the nature of reality, true general laws instantiated by what *has* happened are determinative of what *can* happen, and attention to them is always strongly predictive of what *will* happen. This is so because the basically real is thought of as a world of objects and forces located in space-time, a world which holds together wholly impersonally and not just as seen or grasped from some unifying perspective, and which therefore must be organized exclusively and exhaustively by causal connections. For impersonally and objectively – that is, removed from any notion of an informing intention or shaping will – the *only* possible reason for anything's happening is that something else caused it to happen. These assumptions, coupled culturally with growing technical capacity and declining religious belief, provided the underpinning for the scientific world view as it has been articulated and developed over the past three centuries. They are also (and of course, relatedly) the grounding for our deeply held common-sense conviction that *you can't ultimately buck the odds* – that it is unrealistic (it ignores how things really work) to expect more from the future than you are licensed to expect by the pattern of probabilities derived from scrupulously careful observation of the past. From this disillusioning common sense comes the depressing picture which we were tracing in the first chapter of just how intractable our climate plight must now appear.

But if life will not let us abandon hope for a genuinely sustainable human future, if realism is a condition of all hope, and if conceiving the realistic in terms of what is empirically plausible now leaves us no room for hope of that order, it follows that life must insist on our conceiving the realistic differently – which means that it must insist on something to be underpinned by a different metaphysic of reality. That is a very tall order, given how solidly embedded is the one which I have just sketched, the one which not only regulates practical common sense but on which we rely so extensively to keep all the machinery working. It is nevertheless only from appreciating how tall an order it is, that we shall provide ourselves with any prospect of rising to it.

Empirical realism about our climate situation, meanwhile, now confronts us as we have seen with a brutal dilemma: abandon honesty or abandon hope. It is no wonder that so many thoughtful people, uneasily aware of this dilemma but wanting to hang onto that picture of the real, strive to reassure themselves with various kinds of denial. These include the beliefs, each of them fitted with the appropriate set of blinkers, that something like the Paris accord can be made to deliver, or that a robust global 'sustainability' consensus can still be forged, or that even if politics fails, heroic technology will enable us to avoid or mitigate the more drastic consequences indicated in the previous chapter. But all these blinkers are assumed wishfully in the way that I set out there, and without them none of these beliefs remains empirically credible. Paris and the conventional sustainability model we have already glanced at – and the techno-heroic solution so tempting for our kind of civilization is even more starkly revealed as fantasy, once the blinkers are discarded. It demands an investment of faith in negative emissions technologies (so-called NETs) for carbon capture and storage, or for manipulating the atmosphere, deployed at drastic speed and worldwide scale, when in reality these technologies either have not yet been invented ('non-existent' would better cash out the NE of the acronym in many cases), or have not been tested at scale, or would bring with them such huge associated risks of climate and ecological disruption that the most elementary precaution would shun them. These supposed recourses, essentially the stuff of escapist disaster films, are the final refuge of dishonesty and wishful thinking about our plight.

There is, however, another and an altogether wiser approach, for which I believe the recent renewed surge of climate activism led by Extinction Rebellion is beginning to reach – without as yet, fully understanding its implications. Latterly, as the gathering darkness has become unignorable, a kind of response has been emerging which seeks to reclaim the idea of *realism* for an alternative, non-disillusioned stance towards what it nevertheless seeks fully to accept. The keynote of this response sounds very clearly in these words of the American writer Rebecca Solnit, who has been a powerful advocate for it: 'Hope is not about what we expect. It is an embrace of the essential unknowability of the world, of the breaks with the present, the surprises.' Disillusioned empirical observation indeed picks up on what we have come, reluctantly, to expect – the grindingly slow political foot-dragging, the backsliding and the failures of commitment, as well as the widespread public resistance to any but superficial shifts in habits and lifestyles – and projects these forward. But as Joanna Macy and Chris Johnstone point out in the same spirit as Solnit:

> there is also *discontinuous* change ... structures that appear as fixed and solid as the Berlin Wall can collapse or be dismantled in a very short time ... a threshold is crossed. ... There is a jump to a new level, an opening to a new set of possibilities ...

Or, again, from the Extinction Rebellion 'handbook':

> We work to transform our society into one that is compassionate, inclusive, sustainable, equitable and connected. ... We must all learn how to dream again. ... To break down the old ways of thinking and to move beyond our current conception of what is and is not possible.

The key concept here is that of *transforming* society – and it will readily be conceded that, even bracketing compassion and equity for the moment, the range, scale and speed of change needed to negotiate the carbon precipice illustrated in the previous chapter and reach any chance of indefinite sustainability would constitute a complete social transformation. This term is so important to the considerations which follow, and ultimately for any new understanding of realism to which life impels us, that I must set out at this point just what I am taking it to mean.

Transformative possibilities

A transformation is more than merely a change, even a very significant change. Someone can start acting less selfishly, for example, and if she has hitherto been seriously selfish, that can be a significant change in her behaviour. But we should only say that she has undergone a transformation if the structure of her attitudes and the whole place of self in her moral economy, as it were, had so decisively shifted that, perhaps, selfish considerations no longer even present themselves to her, in a way that simply could not have been predicted beforehand. A transformation, in other words – whether in a person, an object or an institution – is a change not just in content or detail, but in the *form*, which means something like the conceptual frame for understanding and (crucially) for expectation within which we locate its subject.

It follows that transformation is of its nature not predictable – we cannot adumbrate it merely by extending forward the incremental history of what has hitherto been the case. To do that is precisely to depend on holding constant the conceptual form within which we have hitherto understood whatever is at issue. We could predict of someone whom we knew well enough that in response to certain painful learning experiences she might become less, or even much less, selfish. But the disappearance of self from her whole horizon of concern is just what we never could have predicted – it could only have come about through some conversion-type experience to which we would have expected, on the grounds of her past behaviour, that her self-absorption would have made her immune. In the climate case, we can perhaps predict a slow – too slow – transition involving more power sourced from renewables and more investment in technologies for carbon capture and storage. But a switch to 100 per cent renewable energy in a decade, carbon rationing, the severe restriction of private car use and of recreational flying and the whole drastic package of associated measures that might just make the 2°C or even 1.5°C target achievable is, similarly, what we could never responsibly predict – and so, on an empirically realist model, could never realistically hope for.

This, however, is just where the approach represented by Solnit and the others turns the argument on its head. Disillusioning empiricism says that hope for a transformation of this order can't meet the condition of realism, because the odds based on the whole of our past experience are so overwhelmingly against its happening. The response is to say that precisely *because* the changes in question would be radically transformative, expectations based on projecting forward past precedents are bound to overlook or minimize their possibility. Transformation cannot be ruled out on the grounds that the odds are against it, however crushingly, because transformation always happens by definition against the odds. And in the epistemic space thus cleared for transformative possibility, hope can be realistically invested against all the odds, because the addition of active hope into the mix can itself make all the difference.

This latter claim is in effect to re-purpose the idea of wishful thinking. Hitherto I have considered this as a temptation against which realists need to keep up their guard, by cultivating a stance of empiricist disillusion in their approach to what may or may not be possible. But we must also recognize kinds of situation in which holding fast to what one wishes for can in itself be an active power in altering possibilities. Thinking which misconstrues what is likely to happen in line with what we should like to happen tends away from reality only to the extent that the desired outcome is something (such as a lottery win or an acceptable calorie-quantum) to the occurrence or non-occurrence of which our attitude of desiring can make no difference. Across a wide range of human affairs, however, the odds for or against some possible happening are not in this way entirely independent of our attitudes towards its possibility. The clearest kind of case here is the well-documented greater chance of remission or recovery for cancer patients who believe against the odds that they are going to get better, than for those who take a less hopeful view. This feature of individual psychosomatic interaction is also, however, capable of being replicated at the collective level, that of interaction between (as it were) the mind and body politic. In considering whether a certain social change is likely to be feasible, the proportion of those directly involved who both believe it to be so and actively press for it to happen is evidently going to be a factor in the probability assessment itself rather than (as with the lottery or the chocolate cake) simply a piece of disconnected information. Thus, if an empirically unlikely or implausible such change manages to become sufficiently widely hoped-for, the very emergence of this hope will already be helping itself towards a greater chance of passing the test of realism.

Hope in such situations, that is to say, can be so far from being merely wishful as to intervene in the odds against its own plausibility. The concept of the tipping point applies not only in climate science but here too: social, economic and behavioural change which is quite implausible to all observers *ex ante* can happen suddenly and startlingly, it is claimed, once a critical mass of people (one study suggests that this means around 3.5 per cent of the relevant population) starts to believe that it can, and to act accordingly. Transformative possibilities are triggered if enough people commit themselves vigorously to enough specific campaigns, resistances and individual reformations; the overall forms of habit and expectation which make such commitment look forlorn can then suddenly change, literally out of all recognition. Writers like Solnit, Naomi Klein and most recently (at the time of writing) George Monbiot document instances of this kind of transformation from the history of political and social movements. There is the case of the Berlin Wall already mentioned – and more widely, of the sudden dam-burst of change between 1989 and 1991 which swept away Soviet control seemingly clamped down immovably on Eastern Europe, and shortly afterwards the Soviet Union itself. As anyone who (like me) can vividly remember that period will confirm, the Wall had seemed as much a permanent fixture as the Cold War itself, and well after Gorbachev had emerged as Soviet leader in 1985, anyone predicting that within just seven years the USSR would have unravelled, or that before then Germany would have been reunified, would have been dismissed as a fantasist. Other examples, less sudden but no less dramatic in transformative extent, include the process which brought Nelson Mandela from nearly thirty years as a political prisoner of the apartheid regime to become the first black President of South Africa in only four years; the seismic cultural shifts in perception of gender difference and sexual orientation (whatever one may think about some of their later results) which have come about over the last few decades, and the longer-term but no less epochal tectonic shift which during the course of the nineteenth century abolished first the slave trade and then chattel slavery itself (an institution as old as history) across the Western world. These, in their varying time-frames, are all cases where something was hugely unlikely to happen right up to the point where its being actively hoped for anyway by enough people actually *made* it happen.

Common to all such sudden shifts, as Klein in particular emphasizes, is a more or less extended period during which small but dedicated groups of people take local-scale or even just symbolic action against apparently overwhelming odds. And of course, this is just the sort of action which today is going on all over the world in opposition to fracking, in pursuit of divestment from fossil-fuel companies, in defence of indigenous people's rights against extraction or pollution, and in resistance to unwarranted corporate power and influence over public environmental agendas wherever it manifests itself. It is their conscious celebration of themselves as actors in this spirit – 'Whatever the chances, we can no longer leave this to others!' – which gives their characteristic élan to the Extinction Rebellion and school strike movements. We cannot know, it is argued, any more than all those whose cumulative small-scale commitments led to past transformations could know *beforehand*, that all this diverse activity will not suddenly produce a tipping point at which the balance of possibilities dramatically shifts.

It is on this basis, and surely on this basis alone, that the major premise of the vicious syllogism, so compelling to the disillusioned eye, can be resisted, and nugatory gains made painfully slowly over the past twenty-five years can cease to be irresistibly depressing evidence for what is going to be possible over the crucial next five or ten. When we allow ourselves, indeed, to think about what might emerge – what in the past sometimes *has* emerged – transformatively from diverse, networked social action in dynamic and volatile contexts, what suddenly starts to look *un*realistic is the idea that we could ever reliably read off projections about the effects of our interventions in such contexts from the course of events up to now.

To simplify – but not unduly – we could capture the contrasting positions thus: from an empirically disillusioned perspective, it is realistic to admit defeat when the very strong likelihood on any past-facing calculation of the odds is that you are going to be defeated. But from the perspective which envisages transformation, it is always realistic to go on fighting whatever the odds thus calculated, because going on fighting can sometimes, and always unpredictably, change the odds – new action in the human world of attitudes, behaviours and institutions has always the potential to extend the limits of what we have hitherto taken to be achievable.

Recognition of transformative possibility clearly has vital implications for the realism of hope. Crucially, a human capacity to break transformatively from the weight of the past offers us somewhere for genuine hope to lodge – that full-hearted hope for the indefinite continuance of some kind of human flourishing which we have seen the realism of disillusion narrowing down and impoverishing. And it is not just that such hope *can* lodge there – it *must*: for hope has itself to be the driving force of the kind of commitment to collaborative persistence which could bring transformation about. It does so as what Macy and Johnstone, already mentioned, specifically label Active Hope. This is a conscious, structured practice of embarking oneself in working towards the realization of what one hopes for, against whatever apparent odds. It is this deliberately practical rather than merely aspirational nature of active hope which keeps it firmly grounded. Macy and Johnstone's presentation codifies in effect the daily disciplines which enable us to support 'hope against hope'. A strong leavening of what might be called spiritual practice is involved; exposing ourselves fully to the pain of loss and damage, reawakening our gratitude for what we still have and building in collaboration with others the networks of mutual support which will enable us to keep resiliently on, however unoptimistic we may periodically (or even mostly) find ourselves feeling, are all vital components. Active hope means accepting that 'we don't limit our choices to the outcomes that seem likely. Instead, we focus on what we truly, deeply long for, and then we proceed'. Just because success is very palpably not guaranteed, 'the process of giving our full attention and effort draws out our aliveness' and empowers us to recognize that there is equally no reason to prime ourselves for certain failure. By 'making friends with uncertainty', as Macy and Johnstone put it, we can make the issue of possibility open-ended. Or as Solnit expresses the same idea: 'Hope locates itself in the premises that we don't know what will happen, and that in the spaciousness of uncertainty is room to act.'

Nor in the climate change case need such determination be based on ignoring, unrealistically, the inescapable consequences of damage already done. We just shouldn't expect foreknowledge, and therefore shouldn't be deterred by disillusioned claims to foreknowledge, of how these consequences will actually take effect. Whatever future seems to loom, the watchword remains 'Say not the struggle nought availeth'. As Solnit, again, compellingly observes: 'Wars will break out, the planet will heat up, species will die out, but how many, how hot and what survives depends on whether we act.'

Not so unprecedented?

Such advocacy of undaunted active hope reads bravely, and perhaps inspiringly – but only the latter, it might seem, if its claim to the mantle of realism can actually be defended. The empirical stance in such matters, after all, takes itself to be realistic in large measure because it guards against an enduring human liability to self-gratifying illusion – and isn't this interpretation of an uncertain future as offering us open-ended possibilities just as likely to be another self-deception? Doesn't it, for instance, lay far too much weight on the merely logical point that our knowledge of the future can never be completely certain? Much knowledge of this kind, after all, can still have inductive grounds quite strong enough to approximate certainty. David Hume famously pointed out that we can't know for sure that the sun will rise tomorrow, because the question whether it will do so, and the question whether the relevant laws of nature, to which we naturally appeal as backing for our settled expectations, will still apply tomorrow, are in fact the same question – but he also insisted that not to take its rising tomorrow as an unquestioned practical given would be ridiculous. Our climate future is no doubt more open than *that*. But the inductive grounds – the reasons, as set out in Chapter 1 and based on all past experience of what has been politically and socio-economically feasible – for concluding that we have missed our chance to avert catastrophe appear nevertheless (to put it at its mildest) very strong indeed. In the face of them, isn't hope invested in transformative possibility, as a motive to unflagging action, going to turn out to be the last, cruellest form of illusion?

This is an objection which the advocates of such hope have sought to meet by challenging the disillusioned version of realism on its own empirical terrain. They too, as we have seen, call on history – and they do so for evidence of past transformations, in the light of which the present odds against us might look less daunting. (Here we have a clear example of the way in which the embedded metaphysical assumption identified at the beginning of the chapter, that you can't ultimately buck the odds, tends to reassert itself against any challenge: if something is to be able to happen, the odds against it *must* be taken to be less than totally overwhelming.) They claim, in effect, that the apparently unprecedented changes which we would have to make in order to have any prospect of mastering the climate emergency are not really so unprecedented after all.

Vulgar versions of this claim have been trending for a while. Andrew Simms of the Rapid Transition Alliance, for instance, has a short video out there on YouTube in which he insists that 'history is full of examples of how we have achieved the seemingly impossible on a very tight time frame'. But the cases to which he then appeals – early railway network expansion, Roosevelt's Hundred Days, Britain in 1940 … – all just *glaringly* emphasize by contrast how utterly different the present challenge is: global in extent, still in many respects invisible and unfelt, insidiously interwoven with every aspect of our lives and habits, and (attempts to cast wicked capitalists in this role notwithstanding) with no enemy in view except ourselves. Simms succeeds, in fact, merely in reinforcing the very lack of precedent for our peril which he offers to dispute.

There are, however, presentations of essentially the same argument which don't shoot themselves quite so ingenuously in the foot, and it is well worth looking at one of these in more detail, because the deep – at bottom, the logical, or at any rate epistemological – difficulty with the argument needs to be brought out and fully acknowledged if the assertion of an alternative realism is to be properly evaluated. I shall therefore focus on an excellent recent example of the genre, offered by the environmental-political journalist George Monbiot in his book *Out of the Wreckage: A New Politics for an Age of Crisis*.

Political prospects are transformed, Monbiot claims persuasively there, when people's minds are seized by a new narrative which makes sudden sense of their experience in a way that older, failing narratives have ceased to do. Those failing narratives in our present case are on the one hand neo-liberalism, which has produced obscene inequalities, the attempted marketization of everything and chronic financial crisis, and on the other hand a reheated Keynesian growthism which, even if it managed to avoid the high inflation which dogged it in the 1970s, could only now operate by continuing to trash the planet's already groaning ecosystems. That analysis is, so far, convincing in at any rate a broad-brush way. The fresh new story assembled by Monbiot to confront these exhausted options is then that of a huge silent majority of human beings who are by nature 'socially minded, empathetic and altruistic', but whose strong latent desire for a just and environmentally responsible world is currently thwarted by a minority global elite using 'lies and distraction and confusion' as their weapons. Mobilize this silent majority with a compelling narrative of the repossession of our common wealth via a renewal of community and local belonging, he contends, and 'there is nothing this small minority can do to stop us'.

The important point just now is not whether this story is really so new. (In fact it seems to be a combination of Shelley at his shrillest – 'Ye are many, they are few!' – with an update on Kropotkin's *Mutual Aid*.) For Monbiot devotes much space to what he presents as scientific, historical and recent psephological evidence (that is, empirical evidence in each case) for the feasibility of such change – and it is the weakness of this alleged evidence, on any critical scrutiny, which is really interesting in the present context.

Much the weakest is his appeal to science. He cites an article from a (single) psychology journal which identifies human beings as 'spectacularly unusual when compared to other animals' on the grounds of our 'astonishing degree of altruism, unparalleled sensitivity to the needs of others and unique level of concern about their welfare' – a conclusion which is alleged to be confirmed also by studies in neuroscience and evolutionary biology. Countervailing evidence, far from difficult to access, that we are (for example) the only animal which deceives, betrays or systematically tortures others of its own species, is simply not adduced. But in any event, it must surely be obvious that science, whatever its findings, cannot have anything like a final say on such matters, as against the full weight of the humane studies, and (in particular) a literary tradition going back to the Greeks showing decisively that human nature is tragically complex, Janus-faced and morally ambivalent, especially as regards care for one another's welfare. One must assume, indeed, that Monbiot is at least tacitly aware of this, since his references to the 'scientific evidence' in question are frankly perfunctory, occupying less than three pages out of a book of nearly two hundred.

His other empirical appeals, however, are both historical and much more plausible *prima facie*, and it is these which raise the crucial issues here. The more general appeal is to the way that transformational change has characteristically happened hitherto: 'The great emancipations – from women's suffrage to Civil Rights, to independence from empire and the end of apartheid – came about through the mass mobilisation of citizens.' More specifically, he lays great emphasis on the example of what he calls Big Organization, demonstrated in US Senator Bernie Sanders' 2016 Democratic Presidential nomination campaign, as a way of giving such mobilization maximal political effect. This was achieved in the Sanders campaign by a networking process, invented on the hoof, in which volunteer organization was relied on to generate, exponentially, batteries of further volunteers each committed to speaking directly to a certain number of voters on the candidate's behalf. Inspired by Sanders' radical platform of income redistribution, universal healthcare and real action on climate change, a hundred thousand of these volunteers eventually managed to make contact with seventy-five million people – discovering in the process (perhaps unsurprisingly) that real conversations between activists and their fellow citizens can change deeply held attitudes on controversial issues in ways which posters, television advertisements, anonymized mailshots or phone messages from a robot caller are much less effective at doing. The upshot was that Sanders, starting as a rank outsider, almost secured the nomination, and did so without having to sell out to the bland equivocations and big money which enabled the establishment candidate Hillary Clinton to win it narrowly, but then disabled her disastrously in the election against Trump. Had the Big Organization technique been in place from the start, Monbiot suggests on the basis of an insider account of it, both those results might well have been reversed. This realization gives us, he goes on:

> an idea that can be adapted to any situation; a means of mobilising a rolling and self-generating wave of volunteers whose passion proliferates into even wider networks. To read [the insider account] is to release yourself from the poverty of imagination which has locked us into despair. It is to start imagining how campaigns of any kind – not just to win elections but to win the battle over climate change … – can be transformed.

Monbiot writes with eloquent passion. But he makes no move to interrogate his analogies here, any more than he questions the alleged authority of his science – whereas, if he wants these analogies to support his line of argument, he is committed to interrogating them closely. That is because the thought informing any appeal to history to justify the reality of transformative possibility in the present must run like this: sometimes in the past, what seemed utterly implausible on the basis of experience up to that point has nevertheless actually happened – therefore, that X now seems to us utterly implausible on the basis of our own experience hitherto doesn't mean that it can't or won't happen. Or, as Solnit puts this: 'Studying the record … leads us … to expect to be astonished.' A thought with this structure, however, needs to be handled carefully. You can argue that sometimes (in the past) what was then the past has been no reliable guide to what was then the future, and so what is now the past may sometimes similarly fail to be a reliable guide to what is now the future. But you can only do so on the assumption that *generally* the past *is* a reliable guide to the future – otherwise, of course, past astonishments wouldn't license us to expect future ones. And that general empirical expectation of predictive reliability commits us by the same token to identifying criteria of exceptionality if we want to make claims about particular occasions when it might fail. We then have to say something like: in past situations with broad characteristics *a*, *b*, *c* … , the then past provided no reliable guide to the then future; our present situation vis-à-vis climate change has those characteristics; so it is a case in which we could reasonably 'expect to be astonished'.

This reasoning, however, does depend on interrogating one's analogies – in just the way that Monbiot avoids (as indeed do all those writing in this vein). And he does so advisedly, since not avoiding it would require close attention to *how like the past* in these respects the present situation actually is. That would have meant asking whether any of his large-scale examples of successful transformative change really offered a convincing precedent for overcoming the challenges now facing us. These challenges certainly appear utterly novel: we are failing to deal with climate change largely because humanity has never before confronted a danger remotely like this – on this scale, of this complexity, and calling for this order of collective and personal insight, political flexibility, technical intelligence and cooperative restraint. And does the ending of apartheid, for instance – the remedying of a very visible injustice by a state which had been under international economic as well as moral pressure to do so for several decades – really have much to tell us about what will be involved in remedying a pattern of effective indifference towards not-yet-existent future people which everyone worldwide has long been under powerful economic pressure to keep firmly in place? Again, could waves of volunteers, even in their hundreds of thousands, persuade people into mould-breaking lifestyle change anything like as easily as they could capture them for the simple, slogan-friendly objectives – 'Break up the big banks', and so on – dictated to the Sanders campaign by the nature of mass-democratic politics? Different judgements are of course possible in answer to these and similar questions. But making any judgement would have to involve weighing the force of the relevant analogy in relation to our current challenges and identifying the corresponding probabilities, and in any such assessment the probabilities as they present themselves to the eye of disillusioned realism would be bound to exert an enormous – surely, in fact, an irresistible – pull.

If, however, the response to *that* recognition is a further appeal to the ever-present possibility of unpredictable transformation in the face of any apparent unlikelihood, then it becomes plain that the historical analogies adduced as precedents are not doing the kind of justificatory work for which they are being deployed. Those who urge us to invest ourselves in active hope on the grounds that transformative change is always an open possibility can't have it both ways. They can either rely on analogies from experience of past such change, with all their disillusioning potential when we genuinely interrogate their bearing on our present plight – or they can treat what these analogies suggest as always itself unpredictably liable to transformation. What they can't do is rely on the analogies to *show* that this is always going to be the case.

Counter-empirical hope

The recognition towards which these considerations lead us is surely that the hope now needed cannot have the kind of basis for its underlying realism which our long-standing cultural attachment to empirical demonstration pushes us to claim for it. That thought receives confirmation from an initially rather surprising source. Consider this passage by the distinguished climate scientist Kevin Anderson, to whose work I have already referred in the previous chapter, at the conclusion of the article there cited (the full title of which is: 'Climate change going beyond dangerous – brutal numbers and tenuous hope'):

> The current political and economic framework … seems to make [avoiding a breach of the critical 2°C threshold] impossible. But, it is not *absolutely* impossible. If the 'few per cent' of the population responsible for the bulk of global emissions are prepared to make the necessary changes in behavioural and consumption patterns, coupled with the technical adjustments we can now make and the implementation of new technologies (such as low carbon energy supplies), there is still an outside possibility of keeping to 2°C.

That is a very big *If*, when you consider what those necessary changes would have to involve. Yes, as Al Gore and many others remind us, the daily influx of solar energy is enough to power the entire world economy many times over, and the technologies to capture and deploy this energy are rapidly becoming cheaper and more accessible. But Anderson has already identified those 'few per cent' of the world's population, the principal carbon emitters, as 'Large proportions of those residing in OECD countries. Anyone who gets on a plane once a year'. ('Most academics', he tartly adds.) This disproportionately influential minority's making 'the necessary changes' clearly isn't going to be a matter just of a long-delayed *mea culpa* from a lot of overprivileged individuals, but of the very extensive structural and political alterations already canvassed in the previous chapter. Just to recap, we should be looking at a decisive shift away from neo-liberal global capitalism, very significant localization of power to communities and away from global elites, widespread abandonment of consumerist lifestyles and expectations, and massive reductions in the energy-profligate transit of people and goods. Accompanying all that would have to be a shift to virtually exclusive reliance on renewable energy sources worldwide, by not long after mid-century, and a controlled roll-out of whatever negative emissions technologies are found to be both feasible and safe – while *not* using these latter developments as a justification for resiling from any other of these requirements.

Now Anderson, though a powerful polemicist in this field, is first and foremost a research scientist. His whole intellectual formation will therefore emphasize the basing of projections and expectations firmly on carefully observed empirical evidence. But any application of that approach and method to the economic, institutional and sociological grounds for predicting change of the order of magnitude just sketched within anything like the coming decade, which is the relevant timescale, must surely have led him straight to disillusioned realism. And yet he is still, at the end of a paper full of avowedly brutal numbers and their utterly unaccommodating implications, hoping for the 'outside possibility of keeping to 2°C'.

With this example before us, I think we may fairly describe hope so entertained, and in general the kind of active hope that is premised on transformative potential, as *counter-empirical*. It is so in the sense that, however it inclines us to view the historical record, it doesn't really depend on the details of that record for its refusal to be daunted by the odds which a disillusioned realism confronts. This is clear when we see that Anderson doesn't ask, as one would have expected from a scientist of his calibre – or indeed, from any scientist – just what numerical probability the 'necessary changes' carry when weighed against the embedded recalcitrance of our current plight as it appears from all the other brutal numbers to which he appeals. Rather he is insisting, if not quite in so many words and perhaps without consciously intending to (for if his intention had become conscious, his scientific training might well have rejected it), that in human affairs observed regularities, even those underpinned by laws of nature established on the same empirical grounds, *don't always give us rules of probability or possibility for the future.*

It is in this sense that we must understand the appeal to past transformations made by Monbiot and other writers in the same vein. The otherwise despairing recognition that humanity has never before faced a challenge like the one we now confront in the climate emergency is met with the point that the same could have been said of those rallying for freedom in

Soviet Eastern Europe, or of those seeking to abolish the slave trade: their situation, too, appeared to face them with unprecedented difficulties – which we know that, in the event, they overcame. But this is not to rely, as these writers want to suppose, on any argument to the effect that our present situation isn't so dauntingly unprecedented after all. While indeed both we and those past campaigners would speak truly in saying "We've never faced anything like this before", the disanalogies between what the demonstrative *this* picks out in the respective cases remain, as we have seen, so huge as immediately to reinstate the daunting lack of precedent. What is potentially most relevant to our present plight in those past challenges is not so much their demonstration that the then past offered no reliable guide to what was then possible, as the prompting towards *deciding to make it so again*. The life-hope which must inform that decision expresses what we might well, but for the unhelpful religious overtones, call faith in the transformative power of human radical agency. (Faith, as Tolstoy says in *A Confession*, is 'the strength of life' – expressing itself, in a secular context, in the conviction that radical agency is what we really, ultimately have.) That such life-hope *can* triumph is something which the past can encouragingly suggest – there have indeed been these situations in which trust in transformation has been vindicated – but it is not something which past experience can *confirm* for our present case, without our having already committed to the necessary kind of hoping in the present. Commitment means refusing to rule out a transformative leap to a new set of circumstances, what*ever* may have happened up to now and however a disillusioned consideration of past experience might stack the odds against us. Summoning our courage, we determine to act in ways that keep counter-empirical possibility open.

Summary – and back to realism

To summarize: empirical realism, as we saw in Chapter 1, makes the order of behavioural, social and political transformation which we now need seem wildly implausible. But actually, we can't rule out transformation – which is *by definition* change happening against heavy odds – on empirical grounds for any situation where change is not physically impossible. We also know that there are situations in which hope against the odds, that is hope invested in transformative possibility, can create the conditions for its own fulfilment – where hope against the odds can shift the balance of the odds, helping itself to meet the condition of realism just by being genuinely entertained. These are classically situations in which something within the scope of human action was hugely unlikely to happen until the point when its having been widely and actively hoped for in the teeth of that unlikelihood brought it unpredictably to pass. It follows that hope directed at the sort of transformation called for by the climate crisis could then reasonably be held to be realistic if we could reasonably claim the climate crisis to be analogous to other situations of that kind.

The trouble is, however, that we can't support such a claim on empirical grounds – we can't say "Just look at all these encouraging precedents for supposedly unprecedented change here!" – because the odds (as depressingly set out in Chapter 1) against change's happening at a scale and speed sufficient to meet the climate emergency are also odds against any such analogy's carrying weight – that is, against this situation's really turning out to be another one of that kind (and so, *not* unprecedented). These are really just two equivalent ways of looking at what the empirical evidence tells us. That means that if we hold onto the hope that this situation might so turn out, we have to do so counter-empirically, rather than on the basis that there are reliable precedents for its doing so, even against the odds, just here. That in turn means committing oneself to the *creative power* of hope as really adequate to transforming our situation in the face of crisis, a commitment which nevertheless can't finally be warranted by any empirical evidence because it always still finds itself called on in the face of the relevant evidence.

It is hope of this kind which we can perhaps see now emerging to potential awareness of itself in some of the actions and pronouncements of Extinction Rebellion (XR) and kindred movements. Certainly the language in which XR expresses its vision is sometimes closely consonant with the discourse of life-energy impelling transformative hope, as we have been considering this:

> Every crisis contains the possibility of transformation. … We catch glimpses of a new world of love, respect and regeneration, where we have restored the intricate web of all life. It's a future that's inside us all. … And so we rebel for this, calling in joy, creativity and beauty.

And equally certainly, we hazard everything if we dismiss that rhetoric as merely naïve or exaggerated – for without its eagerness and passion, there will surely be no bucking the odds against us. But, by the same token, the danger to which it exposes us must not be minimized. For unless such hope is supported and toughened by full recognition of its counter-empirical status, and is at the same time held *realistically* – that is, as part of holding firmly to the reality of the world in which we have to operate – it is at huge risk of crumbling as time shortens, alleged precedents are exposed as misguided and misapplied, and the odds mercilessly tighten. And humanity simply cannot afford that crumbling. These new, eleventh-hour movements must be able to rely confidently and resolutely on their hopes in the extraordinarily testing sociopolitical circumstances which they will very soon confront – or we are already lost.

This being so, however, we have to recognize that while hope for life's indefinite sustainability comes naturally along with life, whatever the odds, that is still to leave hanging the issue of its realism. That such hope insists on itself as long as we are alive is one thing; whether or not it points to real possibilities is regrettably another – for it is plainly imaginable that life-hope should persist while there remained living individuals, in a situation where in reality the human species had run out of road. That might offer the kind of last-page comfort traded on by some recent novels on the theme, but it is not going to comfort anyone experiencing it in the flesh.

I therefore turn to focus in the next two chapters on how we might indeed see counter-empirical life-hope – creative hope – as genuinely realistic.

3

Creating Possibility

I have been exploring a particular kind of hope, specifically called forth by our current climate and ecological plight. That hope is addressed to ways in which the life-threatening scenario now looming for humanity and the biosphere might nevertheless still be prevented from unfolding at its most drastic. In the course of the discussion so far I have characterized it in two different ways – as *life-hope* in the Introduction and first chapter, and as *counter-empirical* hope in the second. Each of these terms represents a distinct perspective on the nature of this literally vital force.

Life-hope characterizes the hope which we need from the perspective of its natural relation to the instinctual drive of life-energy in human beings. Thought of in this way, such hope is an expression of that drive as it comes to consciousness in a reflective creature endowed with language and reason and aware both of its individual future, and of its involvement in species-continuity through the lives of its descendants. It arises unprompted in the ordinarily robust, healthy individual. As such, it can manifest itself – as frequently in art – in the form simply of an eager openness to the vibrancy of ongoing life, taking no intentional object. But, as brought to bear on the actualities of our present plight, it spontaneously invests itself in the indefinite sustainability of a sufficiently flourishing human life.

When such hope is characterized as *counter-empirical*, however, that is to attend to it from an epistemic perspective. We thereby foreground its ultimate independence of whatever we might have learnt from experience about the scope and tenacity of the obstacles which it confronts and the possibilities of adequate action to remove or circumvent them. In doing so, we emphasize the natural inclination of life-hope to persist against even huge odds in a wide variety of life-threatening situations. And a corollary of that persistence is its investment in open-ended transformative possibility, occurring by definition contrary to the odds.

Taking these two aspects together, we can see how life-hope held onto counter-empirically is a fundamental given of human life, a feature of our way of being not resting for its warrant on any other such features.

As counter-empirical, it does not rest on the observation of regularities determining what is and isn't realistically possible. But nor does it rest on other dimensions of our experience like love, care or imagination, though it can draw important nourishment and consolidation from them; for it does so only in hope that these life-powers can be relied on, and that is finally a matter of hope relying on nothing but its own creative energy. And the term creative here needs to be given its full weight. Such hope, we are strongly inclined to feel, does not just respond to configurations of possibility shaped by the objective (material and psychological) conditions of our lives; it actively generates possibilities in radical agency.

But what is involved in recognizing such generative creativity as a feature of human action? Is it being invoked, for instance, when Solnit calls hope 'an embrace of the essential unknowability of the world', or is that just an epistemic claim about what we can know as the basis for hoping? If it is more than epistemic, what *in reality* does it envisage? Can we, out of our now long-established cultural attachment to a scientific, or at any rate a scientized, conception of the real, make sense of the creativity of hope? If, in the words of my epigraph from Shelley, there really are times when 'Hope *creates* / From its own wreck the thing it contemplates', what does that actually come to?

That is a question on which the whole of my subsequent argument turns. To address it, we need to examine more closely the various ways already glanced at in which people typically surprise themselves by not giving up in the face of huge odds and managing thereby to turn the odds around. If we speak of such situations, as we might naturally do, as occasions of our creating possibility through active hoping, what do we really mean?

Hoping against hope

That form of hope is actually a fairly widespread phenomenon of human activity. I have already mentioned the cancer patient whose confidence in improbable recovery can help her to recover. A similar pattern can be seen in, for instance, wartime and conflict situations where defiance in the face of apparently overwhelming odds helps empower the capacity to justify itself by effective resistance or counter-attack; in the kinds of personal relationship where not giving up on someone can contribute to their becoming, after all, worth persisting with; and generally, wherever commitment which the prospects don't warrant turns out to be fruitful in ways which we simply couldn't have predicted. The common feature in these kinds of case seems to be that hope is deployed in conscious defiance of its own failure to meet what I have called the condition of realism. As in some Eastern martial art, hope turns the very weight of the adverse odds *against* the defeat which they make so likely. This is a process of positive engagement in a way that merely wishful thinking characteristically isn't. Wishful thinking protects itself either by misrepresenting the odds in the first place, as less adverse than they would appear to a clear-eyed empirical realism; or by shutting out all thought of the odds and just concentrating on what it would like to happen. Actively to hope against hope, by contrast, is to face the full adversity of the empirical odds and thereby give oneself at least a chance of leveraging the conditions out of which these odds emerge.

What is going on here can still be captured, however, in terms of the general account of hope which we considered earlier. When I hope against hope, I recognize that my hopes for some outcome X do not meet the condition of realism, since the occurrence of X is just too empirically implausible, but I continue to act *as if* I hoped for X, in the second-order hope that doing so may affect, unpredictably, the conditions which ruled out first-order hope. This second-order hope in its turn meets the condition of realism – though now we have to say, of transformative realism – in kinds of case where refusal to abandon first-order hope at least bears relevantly on those conditions. Thus simply 'believing that something with almost no chance of happening might ultimately happen' (to quote one misconceived psychological account of hoping against hope), and then acting on that belief – hoping, perhaps, that the mugger will have a change of heart and post me back my wallet with its contents intact, and so taking no action to notify the bank of card loss – is really nothing but wishful thinking, because my 'hoping' is evidently just a wheel turning idly here: it could not make any difference, however unpredictably, to the state of my erstwhile assailant's conscience. Equally, making firm holiday plans in the conviction that this time around you will win the Lottery is just wishful thinking, since someone's chance of winning the lottery, on each separate occasion of playing, is infinitesimally small whatever her attitude to playing – whether she just buys the ticket routinely or is led on by eagerly embracing the Gambler's Fallacy can make no difference at all. But in the kinds of case in which we are interested, carrying on as if in hope can produce action in defiance of the odds which is thereby in touch with prospects of transformation.

At the personal level, for instance, your chances of getting to first base with someone whom you fancy may be effectively nil for various good reasons: nobody sensible would bet on those chances, including you. But if you try to 'make your own luck' here, hoping against hope in order to stay in the game, the chances may come to feel just that bit more accommodating – which could make a key difference to relevant interactions, impressions and opportunities. For a more dramatic example involving collective action, consider 1940 (one of Simms' examples of transformative precedent, of course). Realistically, the overwhelming likelihood after the fall of France was that Germany would obtain air superiority to cover a successful invasion of the British mainland (and the air war did indeed come to the balance of a hair). But Churchill's capacity for hoping against hope ("I have myself full confidence ...", as he famously and disingenuously insisted in pledging everyone to fight on the beaches), changed the lights by persuading himself, dissuading high-placed waverers, winning transatlantic admiration which played strongly and materially in Britain's favour, and also of course by inspiring responsive hope in the nation at large – all factors to which the attitude of a pugnacious, rhetorically gifted and effectively empowered leader was clearly very relevant.

Importantly, too, hoping against hope is distinguished from wishful thinking by this kind of being in touch with unpredictable transformation, even in cases where in the event it comes to nothing. The rebel leader who dies on the scaffold hoping that this will not mean the end of his project is not wholly defeated even if his execution does in fact mean that – for his example might have spoken more loudly – while the man who dies hoping until the axe actually falls that an angel will descend to halt the proceedings does not achieve even this qualified triumph over his fate. But although that demonstration of the life-depth of hope even at the point of death would matter for a full picture, what we are interested in here are cases of successfully counter-empirical hope. The issue on which we need to focus is then: given that a condition of realism is still in some sense met in such cases, in virtue as it must seem of an appeal to sinews of possibility which defiant action unexpectedly flexes, what can it mean to speak of that action as *creating* that possibility?

Possibility and human action

In order to press this question, consider a different example from ordinary life. Suppose someone says "I just can't overcome the deep-seated anxiety which prevents me from speaking in public", and is encouragingly told "You don't know what you can do until you try!" (It is revealing that this expresses a very recognizable kind of avuncular folk-wisdom.) Thus encouraged, she does make a huge effort of self-overcoming and finds that she can, after all, speak out at the AGM or whatever. But what was the case with her earlier assertion of impossibility when it was made? Don't we have to say that it was false? "I just can't ..." in such a context must be taken to include the implication that I never will, and this we know to have turned out false; so by simple logic it must already have been false that she just couldn't when she said so. At that point, on this account, there was indeed the possibility of her one day speaking in public, despite her then being quite unable to envisage it. She did not, indeed, know that this possibility existed until her effort uncovered it, but what could it mean to say that her effort in response to encouragement actually *created* it? Surely the most she brought into being here was her own recognition of what was always there *in potentia*? And yet this way of picturing the situation seems unacceptably to underplay her achievement.

Or, to bring this to a sharper focus, and one much nearer to the climate emergency: did Britain in, say, March 1940 with Chamberlain still in office, have the possibility of deterring a German invasion, if Hitler should come to control the Channel coastline? Again, it seems as if it must have been possible at that point for Churchill to have taken over shortly afterwards and, through hoping against hope, to have rallied the country, since that is what did actually happen – the actual must always also be the possible, for if impossible it couldn't have become actual; and then it readily seems too that it must have been possible all along. It must have been, as we might say, *on the cards* even under a weak-willed and uninspired administration that a successful deterrent effort could indeed be mounted – the country must have had the reserves of morale and *materiel* waiting to be called on – so that although it took the accession to power of someone like Churchill to spring that possibility, we have again to talk here about realizing potential rather than creating possibility. And correspondingly, we have to think about this potential as something which would still have been there even had Churchill perished in April 1940 from the coronary which his lifestyle was always courting, and never become Prime Minister at all, so that invasion did in fact follow. But with this thought it also becomes clear that the standing possibility of deterrence must have been completely Churchill-shaped, as it were – only to be sprung through the advent of a leader *exactly* like Churchill – and this now seems to be cheating with the idea of an objectively unrealized potential, a fact of the matter about standing possibility.

These difficulties bring us to the edge of the crucial recognition. When we say that human action can create possibility, the truth to which we are pointing is one which in the contemporary cultural context can be remarkably hard to accept, but which is not for all that especially elusive. It is the truth that in some of our action's most characteristic manifestations, there is prior to action *no such fact of the matter.*

Contrast, for instance, the claim that it could rain tomorrow. We do not need any excursus into the metaphysics of possibility (a morass where one can easily sink) to make perfectly good sense of there being relevant facts of the matter here. Its raining tomorrow is an event which would be consistent with a certain range of meteorological conditions M1, but wildly inconsistent with an alternative range M2 – a settled and extensive area of cloudless high pressure, for instance. (The combination of M2 with 'It will rain tomorrow' will not for all that entail a strict contradiction, that being the standout difference between empirical and logical impossibility.) When we say that its raining tomorrow is possible, we imply that conditions within M1, rather than M2, obtain, and this is indeed to imply ordinarily respectable matters of fact: we are asserting that rain tomorrow is on the cards, because although we are not yet in a position to say whether it will or it won't, its doing so is at any rate not precluded by any of the states of affairs which do overwhelmingly preclude rain on the following day.

The belief that the conditions applying to human behaviour and action, individual or collective, must be essentially like meteorological conditions, only a great deal more complex, is then responsible for the feeling that there must be a corresponding fact of the matter about the anxiety-ridden person's possibility of public speaking or about Britain's pre-Churchillian possibility of deterring invasion. Again, we feel it must have been on the cards all along that invasion could be deterred – or, as we might alternatively say, she must have *had it in her* all along to overcome her anxiety. But that is a deep mistake.

How it can be so may be most readily brought out by attending to the metaphor which we have just re-employed. When we do this, it is not hard to see that 'what is on the cards' in such cases is never complete until it includes the response of the considering agent *to* what is on the cards – which response, of course, may only emerge in action. This is a feature of the antecedents of possible events which occurs nowhere in the world but where reflectively conscious agency is involved. Another way to put it would be to say that the agent's response to the cards must always count as *among* the cards, so that the 'facts of possibility' which the cards constitute prior to action must always be open-ended. But an open-ended fact of the matter does not succeed in being factual at all, since it belongs essentially to the facts (though frequently not to our knowledge of them) that they are *closed* – a requirement recognized since ancient times as the Law of Excluded Middle: any fact either obtains or it doesn't, full stop. In cases like these examples we not only don't but also can't know what is possible for us until we try, because there are no determinative facts about what *is* possible for us before we do try. The anxious person's attitude to her disability expressed in her trying anyway is decisive: it can't have been true beforehand that she couldn't speak in public because in the event she did, but nor was it true that she could – until, by hoping against hope, she had *made* it true. The idea of a prior fact of the matter, straightforwardly explanatory in the case of the rain, wholly misrepresents what is going on here – and that is the substantive sense in which our trying against all the odds does indeed, where it succeeds, *create* possibility. (It follows too, of course, that in such cases there never really are any *odds*, in the sense of objective grounds for a wager placed, as it were, from nowhere. Engaged reaction to what are perceived as the odds must be factored into their calculation, thus always again open-endedly; as at the bookie's, the odds are always evolving in response to people's responses to them.)

Recognition that human action figures thus uniquely among the happenings through which the world moves forward runs, as I have noted, very strongly counter to our scientistically framed assumptions. But so it nevertheless is, and perfectly naturally. Our active creation of possibility in these sorts of case (non-standard, but not actually as rare in experience as all that) brings out especially sharply the very general feature of human action that we are involved in our own possibilities in a way in which nothing is involved in the possibilities of bare events. Our being conscious of our circumstances *as* our circumstances, and being conscious of being thus conscious, constitutes the perspective from within our conditions to which our *having options* has to be referred in order to be properly understood. The possibility of its raining tomorrow comes down to its being neither precluded by the relevant circumstances nor clear yet which way things will pan out; the possibility of my going for a bike ride tomorrow depends in part, indeed, on my being neither bikeless nor legless, but otherwise turns on my not having decided yet. It would not only be very strange, but would evince psychotic inner dissociation, if I said (as about the rain) that it is not yet clear which way things will pan out as regards this biking, so that I must watch for developments; and this awareness of not having to wait for the facts to declare themselves is my awareness of being actively *inside* the coming of those facts to pass – of my decision's *making things happen*, rather than being itself just one more happening.

Action and causality

Here, indeed, we can feel the metaphysical ground of empirical realism beginning to shift beneath our feet – and to shift under seismic pressure from our life-need to find counter-empirical hoping realistic. For that ground consists in the fundamental assumption that human action is governed like everything else in nature by causal laws which predetermine what is objectively possible in any given situation. And what is suddenly no longer firm, when we think closely about the implications of hoping against hope, is the part of that assumed terrain on which the ancient philosophical conundrum of free will has always been located. About this, even in a book which seeks to minimize explicitly metaphysical discussion, I cannot avoid briefly saying something.

The conundrum, familiarly enough, goes as follows: in order to be responsible for my actions, and especially to be morally responsible, I must have freely chosen to perform them. That is, it must have been true that when I did so choose, I was fully able to have chosen differently, so that my having or not having done so was wholly in my hands or on my account. But if I am naturally embedded in a world organized through and through by causal laws, this can never be the case. For in such a world, nothing which actually happens could have happened otherwise than it did, or could have failed to happen – that being at the core of what causality means. (If an event is caused, then once its cause is in place, it *has* to happen, and lacking a cause no event *can* happen – deny either of these claims and not only all explicability but all coherent connection between events immediately dissolves.) And surely I am indeed so embedded: my choices, expressing my physical and mental constitution and history, are natural events in entirely the sense to which universal causality (the principle just enunciated) seems to apply. So it seems that this universality of the causal on the one hand, and human freedom on the other, are incompatible, and we must give up on one or the other – but both seem to be indispensable aspects of what we take to be real.

Since Hume, the default attempt to prevent this puzzle from destabilizing the metaphysical ground of empirical realism has been acceptance of a position called compatibilism – short, as it were, for *non-in*compatibilism. On this understanding of the issues, there can often arise the situation where someone's choices are unconstrained by factors external to him. Since he is not being forced to act by anyone or anything *else*, he can truly think of himself as acting *freely*, in that sense. But this remains perfectly consistent or compatible with his choices' being determined in such situations by his own history, conditioning, habits and so forth, which together cause him – in the ordinary determinative way of causality – to give weight to some reasons over others in considering how he will act. S acts freely, on this model, to the extent that S is determined only by *his own* causality working through him.

Unsurprisingly, that account has provoked reactions ranging from seeing it as simply obvious to rejecting it as blatantly changing the subject. But to say, as in the previous section, that human action is distinctive by virtue of the agent's being uniquely *inside* the events in which his or her choices consist, might easily be conflated with it. They are, however, very importantly different positions. To see how, imagine first a truck moving down a railway line towards a set of points where the line bifurcates. The points will be set to turn it onto one branch or the other, or else they will be positioned (jammed, let us suppose) somewhere in the middle. So the truck either veers to right or left, or it is derailed – having at no stage of the process anything to do but go where it is sent. But now transform this into a picture of yourself consciously approaching a choice to be made between two options. The general analogue here to there being no prior facts about possibility in the cases of hoping against hope which we were considering, is that as you approach such a choice, *nothing need correspond* to the points' already being set one way or the other, or being jammed in the middle. It can be solely your upcoming decision which will do the equivalent of setting or failing to set them. Nor need that decision be predetermined one way or the other by the history and conditioning which you bring with you – if it had to be, that would just return you to the truck's case, with the points already in effect positioned beforehand. Radically unlike the truck, however, you are (and of course, all along know yourself to be) the agency which *produces* the determination of your direction, doing so through your active reviewing of options which remain open until you have made (not 'been constrained to') your choice, by giving weight to (not 'registering the weight of') one set of reasons as against another.

It is crucial to see that this understanding of the matter does not exempt the chooser from general causality – but it does change the vital relations in which he or she stands to it, in a way which it is compatibilism's whole point not to do. At moments of choice, I *implement* my history and conditioning, not as its manifestation but as its agent, actively extending its reach forward to configure each new situation. (And every human choice-situation is always new: a unique individual, never replicated from world's beginning to world's end, confronts a situation which cannot ever be identical with any preceding one since each can be differentiated by its potential to contain a memory of what has gone before.) The causality constituted by my life-conditioning and brought to a determining focus by my considering between options, thus operates through my choices but not *on* them. Just as the current electrocutes me if I touch the live wire, but doesn't ever electrocute *the wire*, so we might say that my unfettered choices are live with (my own) causality, but are not themselves caused. And in that radical distinction from everything else causal resides human freedom generally, as well as our capacity to create possibility.

That this is no form of compatibilism is most clearly evident from the fact that that latter position can be articulated wholly in third-personal terms – as indeed it was in introducing it a couple of paragraphs back. But we can only really understand what is going on in free choice by adopting the first-personal perspective of the choosing agent, to which not-yet-determined options present themselves irresistibly as *still open*, and taking that perspective with fundamental seriousness: that is, treating what our subjectivity reveals to us as unquestionably real.

How could its reality be impugned? Surely it is just self-evident that our subjectivity is *inherently* active and agential? If you try to imagine yourself as a consciousness which rather than deciding its way forward has always to wait passively to find out what action it will become aware of performing at each choice-point, not only will you experience an extreme form of dislocation (as already noted), but actually your sense of being a *self* at all will feel threatened with dissolution. (Try it, now. Then try thinking that your trying it just happened to you.) Our selves do not sit quietly in Daniel Dennett's 'Cartesian theatre', with the contents of our consciousness unrolling on a private screen before us – they are permanently busy, each in what might (in the spirit of the image) be called its Cartesian lorry-cab. So if the irresistible conviction going with that busy-ness, of our being faced pretty often with genuinely open options, were actually an illusion, then our whole sense of subjective selfhood would have to be regarded in the same light – that is, as basically illusory. Now people in thrall to the Scientific World View have indeed tried to argue just that. A particularly egregious example comes from the molecular biologist Francis Crick: '"You", your joys and sorrows, your memories and your ambitions, your sense of personal identity and your freewill, are in fact no more than the behaviour of a vast assembly of nerve cells and their attendant molecules'. Crick calls this 'the Astonishing Hypothesis', though he manages in the book in question to be quite astonishingly unastonished by it. Personhood, for him, right down to the self as subjective centre of perception, memory, will and action at its core, is an evolutionarily useful trick that our neurons play on us in the course of their wholly determined dance – an elaborated illusion helping to promote our survival. Nor is Crick by any means alone in offering such an account – Dennett himself and Thomas Metzinger are only the more illustrious among others who have recently done so.

But subjectivity can't be an illusion: the very concept of illusion *presupposes* subjectivity. An illusion is an appearance taken, thus *by a subject*, to be the case when in fact it is not. So for anything, including subjectivity, to be at bottom an illusion, subjectivity – awareness of oneself as an active consciousness taking appearances as veridical – must be fundamentally real. And certainly if such subjectivity isn't an illusion, it must be real. So it must be real.

At first blush this just looks much too quick. How could an argument so simple and obvious have been overlooked, for instance by the luminaries just cited, unless there was something seriously amiss with it? But actually there is nothing amiss with it, and it has only the kind of obviousness which very easily hides in plain sight, because it strikes so deeply against the ferociously ingrained cultural prejudice underpinning the Scientific World View – the assumption that reality is ultimately third-personal – that it *has* to go unseen. And once that prejudice is shaken by what I have called the force of our life-need, the overwhelming urgency of finding hope-against-hope realistic, then again the metaphysical ground of empirical realism is being put in jeopardy. Third-personally, the world cannot contain undetermined areas, gaps where things are so far neither so nor otherwise. (Controversially, it may contain areas at the quantum level where we cannot in principle *know* one way or the other.) The objective world, third-personally viewed, must in itself be such that either there is going to be sufficient reason for A to happen, when it must happen, or there isn't, when it can't. So the perspective of agential subjectivity, in which there really are such gaps around human action, must from the outside (third-personally) be regarded as a profound illusion. But once we see that it cannot be illusory, the third-personal viewpoint itself must be rejected as constituting our only access to the fundamentally real.

The collective-action dilemma

Registering that the creativity of hope is real in the way that I have outlined, is the single most important step in understanding how its investment in creating transformative possibility can be realistic. And without such understanding we are disarmed in face of the climate emergency – to the direct consideration of which we must now return, because the thoughts in the preceding section only matter in the context of that emergency if recognizing how we can create possibility makes clear sense of, and so empowers, the hoping against hope to which life now calls us in this arena. So I want to examine one of the most characteristic features which our climate plight seems to present, and one which seems most directly to confront the claim that we create possibility – one which presents us, indeed, with our apparently deep-seated *impotence*, both individually and collectively, to do anything remotely adequate about our situation. Taking this together with a recent forceful and well-publicized expression of hope in the face of it, offers a chance to bring these issues of creativity towards some kind of test.

The feature in question is that much of the intractability of the climate emergency lies in the enormous and unprecedented scalar disjunction between human motivation and action on the one hand, and atmospheric reality on the other. Human beings have literally never buggered up anything as big as this before; the changes threatening us do so at the level of whole-Earth systems. As such, they are not just huge – they elude our direct experience and can only be grasped at a theoretical remove. While we can sometimes perceive more local alterations such as the gradual retreat of a glacier or the decline of bird-life in a particular area, changes at this level are not felt to *threaten* us as a climate emergency except when interpreted as manifestations of wider systemic derangement. Even things such as floods and wildfires which do directly threaten us, or some of us, do so as part of such an emergency only under similar interpretation. As a corollary, the motivation for remedial action addressed to underlying systemic causes has to be constructed from attention to the data (or, much more usually, to expert accounts of its significance) as well as to the relevant imperatives – that is, it must be generated through essentially theoretical considerations, rather than arising from immediately felt pressure. We could contrast here the motivation to early socialist action, much of which came from seeing, smelling or otherwise directly suffering the gross poverty which it was hoped that social change would alleviate. It perhaps sounds bloodless to refer to the desire to avert grievous suffering for your children and grandchildren as a theoretical motivation – certainly, as I noted in the Introduction, that desire felt in the presence of one's actual grandchildren is painfully acute – but what remains true is that just to know that their suffering is really coming, you have to do a lot of theory-informed interpretive thinking (something for which many people tend in fact to be poorly equipped).

Meanwhile, those who do pay the right kind of attention are condemned to awareness, as part of the same theorized recognition, that emissions-reducing action by any individual has vanishingly negligible effects, and action by any collective much smaller than the whole world acting in concert would still be insufficient to make any decisive impression on the trends now building towards disaster. So my refusal to fly any more or your abandonment of a lifetime's meat-eating habit, while likely to be big deals for each of us, are in climate terms not even straws in the teeth of a hurricane – and similarly too at much larger scales. Britain's carbon emissions, for instance, if totally halted tomorrow, would only reduce worldwide levels by a helpless 1 per cent – and even cutting out the much larger percentage attributable to the US or China could not in itself halt the overall warming trajectory.

This all contributes to making our climate plight into the mother of all collective-action dilemmas. (For reasons which will become clear in a later chapter, I use this term in preference to the usual 'collective-action *problems*', but the structure in question is the one standardly recognized.) These dilemmas are cases turning on the lack of incentive for some actor to commit as first-mover to a change which contributes so minimally to the establishment of a collective benefit that in itself it makes no difference, but which is initially significantly disadvantageous to the actor in question. Or, putting it the other way around, they turn on the incentive to free-ride when a benefit needs to be achieved for us all by everyone working together, but can equally well be achieved, as far as I am concerned, by everyone *else* working together. Since under these conditions no agent is rationally motivated to change their behaviour before everyone has done so, no one does change, so there is no collective change and the collective benefit fails to accrue. Technically expressed, for each agent defection dominates cooperation as the individually rational action whatever other agents do – if I don't change and others do, I can free-ride, while if I don't and others don't either, at least I am not putting myself fruitlessly at a disadvantage. This generates a dilemma because, on the one horn, each pursuing what they want produces what nobody wants, while on the other, pursuing what everyone would like demands, unpersuasively, that each do what is against their individual interest.

It has long been recognized that this dilemmatic pattern presents itself very clearly in environmental and resource-usage contexts – as such, a version of it was famously labelled 'the tragedy of the commons' by Garret Hardin back in 1968. Each herdsman grazing stock on a common has the incentive to add one more beast, since his doing so cannot in itself overgraze the available area, and by the same token he has no incentive not to, since whatever others do his particular bit of restraint won't save the resource – and consequently everyone does go on adding beasts one by one, until the common, thus overgrazed, is ruined for all. The public good constituted by any natural commons such as a river, for instance, will tend to be abused for reasons structured in just this way: each agent (including, importantly, corporate agents like firms) lacks the motivation to restrict their own damaging use of the river, as a destination for effluent perhaps, before every other agent involved does so, and so no one acts and the river becomes polluted with high costs to all. The same disincentive structure leads to overfishing and other well-recognized forms of natural resource depletion, including of course that of the atmosphere's capacity to regulate planetary temperature within humanly habitable boundaries.

It is also well understood how such situations can be addressed locally in various ways, from long-standing customary restraints to appropriate kinds of regulation. But the structural imbalance between action and effect is, as we have seen, clearly evident at its starkest in the case of the climate, where these options for resolving the dilemma are also much harder to envisage. There is no global regulatory body with the necessary authority and powers of enforcement – and exactly the same dilemma is repeated at the level of disincentives to the establishment of any such transnational authority, or to complying other than nominally with the dictates of whatever poor substitutes do get established, as these disincentives face individual nations. Nor is it easy to see how anything analogous to customary restraint could operate in the global context: custom in the form of folk-wisdom can restrain the over-exploitation of a particular natural resource, but only in the kinds of local, face-to-face economy which globalization has largely overridden – mainly, indeed, through this very structure of disincentives against maintaining traditional restraints. On top of this, of course, has to be factored in the time dimension affecting climate-related action: emissions restraint, for instance, will always be a cost imposed on themselves by present people in the interests of benefitting a cross-temporal collective (humanity, or the living world, as a whole) in which, by the time the major benefits accrue, we ourselves may well no longer be around to participate.

Hence the climate emergency must be understood as constituting the largest and most complex dilemma of this form in history – and this is one important reason why past examples of transformative action against evils such as slavery or unwarranted gender discrimination turn out to be so disappointingly disanalogous to our current plight. As such, essentially the same dilemmatic structure operates at every level, from the householder wondering whether solar panels are going to be worth the effort and expenditure, through the holidaymaker considering transport options, all the way to the nation state finessing its commitments to international agreements like the Paris accords in the knowledge that all the other signatories will be finessing theirs.

'No one is too small to make a difference'

The collective-action dilemma looks very formidable as framed theoretically, and a disillusioned view would recognize entirely predictable results from its shaping of both individual behaviour and policymaking in response to the climate crisis since we first understood that there was one. What would it mean to 'create the possibility' of escape from this dilemma? And we *must* believe that we *can* escape from it, if there is to be any chance of avoiding the forced march to catastrophe which it seems to promise. That has to mean hoping to find our way through, or around, a barrage of difficulties which exhibit the same core structure. As Dale Jamieson has written (though he uses the standard 'problem' terminology),

> climate change does not involve just single, intra- and intergenerational collective action problems. Jurisdictional boundaries and competing scales cause multiple, overlapping and hierarchically embedded [such] problems. A vast array of behaviours by individuals, nations and other entities affect climate, but they are governed by an equally vast array of different regimes with different mandates.

And it follows that in the nature of this case we are going to be hoping against hope, since these nested iterations of the dilemma pile up the odds more and more overwhelmingly against us. To put it in terms of our preceding discussion, the collective-action dilemma seems to spell out the *structural* impossibility of the world's averting climate catastrophe: a structure grounded firmly in the nature both of cumulative atmospheric change and of ordinary human motivation. Here, if anywhere, then, the hope in which we nevertheless go forward must be seen as able to create new kinds of possibility, if it is to be taken as transformatively realistic.

A powerful recent expression of such hope, and one which seems deliberately aimed at the heart of the dilemma, is the claim that 'no one is too small to make a difference', slogan of the remarkable and admirable school strike instigator Greta Thunberg. So what, in the light of the foregoing discussion, might we make of this claim?

The dilemma gets its grip, as we have noted, in virtue of combining two distinguishable elements. The first of these is the disparity of scale between particular action and atmospheric effect. The second is the disjunction between the perceived interests of each rational agent (individual or institutional) and those of all other agents, such that in principle my interests could be served at the expense of those of the whole community, or ours at that of the rest of the world; this might also be labelled the principle of free-riding where you can. (That in the climate case undercutting the common interest so evidently ends up impoverishing everyone further down the line is just the destructive point of the dilemma, since recognizing that to be so actually gives me an even stronger individual interest in free-riding while I still can, which makes longer-term catastrophe all the more predictable and thus gives me … and so on.)

If we now ask where we should be directing the energy of counter-empirical hoping against the grip of this dilemma, it might seem that expressing things in Thunberg's way points towards the issue of disparity of scale. (She herself plainly intends to reference ironically her own slight physical stature in this connection.) But her slogan can't be offered simply as a challenge to the truth that *anyone*, considered individually, is too small to make any significant atmospheric difference – that truth is evidently incontestable. Rather it must point to new avenues for scaling up individual action.

Thunberg, of course, is an internet phenomenon, in the strictest sense – that is, she would not have become a phenomenon at all but for the internet – so perhaps the slogan directs hope towards the online connectivity which now extends the reach of individual example, enabling it to capture the attention and endorsement of millions? And it is certainly true that no one is too small (nor, in notorious political and 'celebrity' cases, too stupid or vacuous) to have a Twitter following. The trouble is that the operation of even genuine influence or inspiration through this medium is still entirely vulnerable to the dilemmatic structure, as long as we think of what people are being influenced towards as action taken for prudential reasons, action in their own longer-term interests. For however many people are inspired by one influencer's example, they all still have to act as individuals for this inspiration to start having any cumulative effect, and each will still be rationally motivated to hesitate until he or she sees how others (who will be similarly hesitating) act on their inspiration. How far each lets a surge of socially mediated incentive carry him or her will thus remain dilemmatically configured. Each of us may do enough to register our enthusiasm, though without turning our lives upside down, and the cumulative result would still be nowhere near making a genuine climate difference – and each, being aware of this, will be the more inclined to compromise with his or her enthusiasm. The keynote of the dilemma, to repeat, is that 'defection dominates cooperation' *whatever others do*, and we see this inexorably replicated even as individual actions begin to build towards a collective wave. Being inspired only *so* far will dominate reconfiguring one's whole way of life in the common cause; taking only limited action to exert pressure will dominate going the whole hog along with a revolutionary crowd; acting to deflect such pressure rather than to make far-reaching systemic changes undermining their own positions will dominate responses by institutions at all levels; and waiting until other states make the really difficult moves will dominate taking a lead and organizing a confederacy of action on the part of national governments. So it looks as if the determination to hope against hope, which like all creative action always has to start with committed individuals, will run into the sand of ineffectuality – or at best, strictly limited effectuality – long before it has the chance of stimulating changes of the extent required.

We must therefore, if we are to make sense of it, rather see the slogan that 'no one is too small to make a difference' as an investment of hope in trying to shift the model of the rational agent itself which underpins the whole dilemmatic pattern. And here the obvious target is the assumption of *prudential self-interest*, the economist's working assumption (or, the assumption which makes economics work) that an agent's rational motivation is towards maximizing advantage – net benefit over cost – accruing to him- or herself. We might then read the claim as meaning something like: 'No one is too small to set an example of morally responsible instead of self-interested behaviour in these matters' – refusing to fly, for instance, however inconveniently to oneself, since even a short-haul flight generates emissions exceeding the annual total of someone in Uganda or Somalia; or protesting in uncomfortable and troublesome ways because somebody has to. This indeed appears to be what Thunberg herself intends when she says things like: 'The bigger your carbon footprint – the bigger your moral duty. The bigger your platform – the bigger your responsibility.' We have here what looks like a new dimension of morality, arising in response to our unprecedented times. When someone comes to the point of simply refusing to be part of a system which is destabilizing the biosphere, extinguishing species and jeopardizing the human habitability of the planet – never mind going on actively to oppose that system – he or she takes up a behavioural stance not only never demanded of any earlier human beings, but never previously available to them. Here then, in its relation to this new form and force of responsibility, must be where the creativity of hoping against hope can genuinely bring into being the possibility of our escaping the collective-action dilemma.

This will turn out to be true, and crucial to our chances – but only once we have properly understood the nature of the responsibility involved. This theme, though essentially continuous with that of creativity, is important enough to deserve a chapter title to itself.

4

Responsibility Beyond Morality

We ended the previous chapter by concluding that the difference which no one is too small to make must be a matter of helping to dislodge the assumption of self-interested action which underpins the collective-action dilemma. Unfortunately, appealing from self-interest to *moral* responsibility fails to escape the essence of this dilemma. The trouble is that the disparity between action and climate effect remains in force to prevent morality from getting any real grip here.

At first blush that seems counter-intuitive – surely the moral issues in this area are starkly clear? Whatever sense it makes in general to attach moral predicates to a collective entity (unsurprisingly, a contested point in social philosophy), it must be taken for the purposes of this book as perfectly intelligible to say that Western-style civilization, in persisting with its high-carbon ways of living while knowing what it ought by now to know about their consequences for climate and biosphere, is acting with gross irresponsibility. Indeed, attributions of climate irresponsibility to anything other than a large enough collective entity would be *un*intelligible, since only of such a collective does the claim that its actions are jeopardizing the climate even make sense. Again, however contested might be the issue of irresponsibility towards *what* – the question of whether an extensionist environmental ethics which talks about duties to threatened species or to the biosphere itself makes sense – it surely cannot be denied that future *people*, at least, are being thereby treated irresponsibly, nor moreover (as emphasized in the Introduction) that some of those who will suffer serious future adversity from such dealings are already alive. And the failure of those in prominent positions, most glaringly the world's so-called political 'leaders', to have led their various constituencies towards accepting the need for change has been a form of dereliction of duty beside which ineptitudes like blundering into war or Brexit pale in comparison. It would seem to follow that there must be a moral imperative for this civilization to introduce constraints on its harmful high-emissions behaviour as a matter of very great urgency; and that looks as if it would translate straightforwardly into a corresponding duty for each one of us both to press vigorously for such a collective response and to constrain the high-emissions aspects of our own personal behaviour meanwhile – with no one being too small to set an example by doing so.

In its broadest outline, this picture must be right. Our hope against hope can only be that people in very rapidly increasing numbers will at last start to act vigorously in both these ways, and that governments, commerce and other collective institutions will either be put under irresistible pressure to respond, or be swept aside in favour of revolutionary alternatives. But modelling these vital processes in terms of specifically *moral* responsibility, at either the collective or the individual level (which has been the default tendency of those wishing to escape the prudential dilemma), is actually likely to confuse and hinder rather than encourage it. Paradoxically, it makes our real obligations here much easier to evade.

Morality and the atmosphere

The essence of morality as we ordinarily understand it is, and always has been, the 'golden rule': don't do things to other people that you wouldn't want done to you. More orotundly: before acting, put yourself in the place of those people whom your contemplated action will affect and consider how you would feel about undergoing rather than causing those impacts. That we recognize as perfectly natural the need to bring our actions under this kind of constraint, rather than just blithely pursuing our own individual interests, reflects at bottom the interdependence which we inherit from primate sociality. A moral obligation is then, in general terms, a duty to behave towards someone in a way which takes their relevant experience thus into consideration. Although this formulation mentions 'impacts', it captures something which is completely common ground for all the main normative ethical systems figuring in Western philosophy, each of them offering a different slant on it: consequentialism claims to provide a notation for tabulating the upshots of the given kind of consideration, Kantian deontology offers a technique for conducting it in a rationally standardized fashion (rather than just depending on one's powers of sympathetic imagination, or not), and the virtues are character traits which both conduce to thinking like this and are developed as part of learning to do so.

The core of the difficulty in the climate case is then that any collective large enough to be chargeable with causing climate-driven impacts cannot engage in anything which might count as 'taking the experience of others affected into consideration', while the kind of agent who *is* able to do this cannot cause relevant impacts. Intermediately, a smaller and more 'governable' collective (such as a nation state) which does have structures enabling it to exercise what we might recognize as collective moral agency, does not produce climate impacts in a determinate enough way for this to translate into genuine moral obligations, either collective or (therefore) bearing indirectly on its individual citizens.

The first point here reprises one already made in the previous chapter and glanced at again earlier – the point that if we are to ascribe current responsibility for the climate emergency at all, it must be to something as huge and amorphous as 'Western-style, urbanised, fossil-fuel-based civilisation worldwide'. Historically, of course, rather more compact human groupings, especially nation states like Britain, the US and those of Western Europe which initiated and carried forward the Industrial Revolution, were principal actors in building up the current concentration of atmospheric carbon dioxide. But what has brought the world to climate crisis is extension of this carbon-based economy to serve a rapidly expanding human population right across the globe. To say that *humanity* owns this crisis no doubt misrepresents, since there are still wide tracts in which people living at or near subsistence level generate emissions only from activities essential to their daily survival. But equally, forms of consumption-driven carbon profligacy are now so general as a lifestyle norm, and so characteristic of the whole tendency of 'development', that the misrepresentation has some excuse. Certainly, the causal nexus is now so far-reaching that even if the presently worst national offenders – the US in terms of per capita and China in terms of total consumption – were both to become carbon-neutral tomorrow, dangerous warming would be merely slowed, not halted or reversed. And correspondingly, the 'collective' which urgently needs to take concerted action if crisis is not to become global catastrophe extends far more widely than any nation state or (crucially) any effectively organized grouping of such states. For a necessary condition of predicating moral responsibility or agency of any *collective* is that it exhibit a degree of integration and organization such that, at the least, functions analogous to those of reflection, consideration and decision which are key to being an *individual* moral agent – that is, something capable of taking the impacts of its behaviour on others into account – can be exercised on its behalf by properly constituted representatives. Nation states with established forms of government and recognized leaders can often meet this condition; the carbon-profligate world community (even if it had relevant others to envisage) palpably cannot, as the helpless gestures of the United Nations in this direction make only too plain.

But then, the thorny issue raised in that last parenthesis must also be acknowledged. Who would be the relevant others in duty to whom the 'present world community', or that very large portion of it which is emitting too much carbon, had a moral obligation to change its ways? The easy response here has been to invoke 'future generations', but only a little scrutiny reveals this as far *too* easy. Duties to as-yet-non-existent people can be nothing more than pseudo-duties, since in principle it can only ever be down to one party concerned (that is, present humanity) to determine how far the obligations supposed to be involved are or are not being met. That, of course, is the hook which the sustainable development paradigm is meant, but fails, to let us all off. (It is as though I owed you money, but had complete discretion over attributing value to the currency – buttons, fir cones, stones … – in which I might opt to pay you back; evidently, the notion of *owing* is doing no real work here.) Nor can we get around this difficulty by substituting 'the presently rising generation' for 'future generations': the world community at any one time significantly *includes* its rising generations for carbon-emitting purposes, and these generations do not in any case form a well-defined collective over against the world's adults, but are unceasingly merging into this latter category on a massive scale. They are not, that is, 'other' enough to count as morally relevant others.

So we might find ourselves driven by the template of moral obligation here to identify as relevant others specific groups of present people, such as the inhabitants of the Maldives for instance, whose contribution to rising emissions has been utterly negligible but whose sufferings (from rising sea levels in this case) are clearly down to global warming driven by most of the rest of us. But then, the shape of a moral obligation has been formally maintained at the price of losing any grip on climate crisis as a whole: for given all the other factors impinging on this kind of situation, all we might have established is merely an obligation on the world community to spend a few billions on relocating and compensating these inhabitants, and anyone else similarly placed. This is something, of course, especially if you live on the Maldives, but it is a very long way from what we wanted to capture with our broad picture of global responsibility to tackle climate emergency.

If the carbon-profligate world community cannot really count as a collective moral agent in respect of the emergency, what about those who unproblematically do count as moral agents – high-consuming individuals? Here if anywhere, surely, and most particularly in appropriately informed people of conscience and goodwill, is where hope against hope for transformative change must be invested – so isn't this where the moral energy to drive such change must, like all human energies, ultimately originate? Unfortunately, however, the template of moral obligation fails even more thoroughly to fit our climate-related activities at the individual level, and this emerges very clearly when we press the question just how any individual's present high-emissions lifestyle can be morally *wrong*, as it must be if he or she is to have a moral duty to change it and can set a moral example by forswearing it. For it follows from the scale disparity already emphasized that any individual's high emissions will themselves generate no climate-systemic effects, and so cannot generate *harmful* such effects. Correspondingly, there is no suffering which, for example, my refusal to take a long-haul flight will prevent – despite the fact that these flights emit measurable quanta of carbon per passenger, which quanta can be summed and the totals correlated with the likelihood of measurable adverse impacts on people living later on. Any rule against such emissions cannot therefore on this model either be, or be grounded in, a consequentialist moral rule.

I have found in discussion that this claim strikes some people as counter-intuitive. Individual long-haul flights, they urge, can be reasonably regarded as contributing directly to suffering – each flight certainly contributes its carbon emissions, and each contribution must have an effect however minimal, otherwise a volume of emissions large enough to cause suffering by altering the climate could never be caused to accumulate. It might then seem obvious that, to quote Ms Thunberg once more, 'the bigger your carbon footprint, the bigger your moral duty' to reduce it. The nature of the climate system, however, means that matters are not that straightforward. What we are doing collectively does indeed cause future harm, and the long-haul flyer contributes his or her mite to what we are doing collectively. But equally clearly, this mite does not itself *cause harm*, and this is not just a matter of the harm it causes not being *identifiable*. It is not as if one were (in an analogy due to the philosopher Derek Parfit) firing off an arrow towards a crowd so distant in space and time that one will never know if anyone got hit. Here, even though in principle I could never find out, it remains the case that if my arrow does hit someone, it was my arrow that struck home. Emissions, however, don't function like that – the atmosphere not only immediately anonymizes them, it also as it were causally disinherits them. A more accurate analogy for the long-haul flying case would be that millions of us fire off arrows, none of which ever in fact hits anyone, but the combined draught of their flight triggers an electrical storm in which someone is struck by lightning. My arrow plainly does not cause this person's death, but nor would *any* other consequence of the electrical storm be prevented by my not loosing it off, and of course the same can be said of everyone else's arrows, individually considered.

The revised analogy is still not perfect, of course, because randomly and pointlessly loosing off a lot of arrows would seem like a careless business to participate in, if it is indeed likely to cause electrical storms, whereas all the various things which people do to generate theoretically dispensable carbon emissions have each their own rationale – even a long-haul flight can, in certain contexts, be undertaken to meet a genuine moral obligation, such as visiting a dying relative. Nor is this like Parfit's other relevant thought-experiment, the Harmless Torturers, where each one of a thousand people turns a dial a thousandth of a revolution, the full revolution subjecting someone to excruciating pain. Some would intuit that each here, while doing no discernible harm individually, errs morally nonetheless. But again, this intuition arises (where it does) because twitching the dial does nothing except contribute, albeit minimally, to torture; also the contribution is direct and each increment is in principle causally attributable even though indiscernible. None of these things is true in the climate case, so an intuition of wrongness there cannot have any real consequentialist basis.

For exactly the same kind of reason, your high emissions cannot be wrong because they are either vicious or unjust. Restraint is not a virtue as such, otherwise restraining oneself from attempting Sudoku puzzles or eating Swiss cheese would be virtuous: it is a virtue where the impulses being restrained are harmful, just as generosity (for example) is a virtue because it disposes us to share our substance with people who suffer from not having enough. But we have just seen that in the climate case the individual emitter causes no harm. Nor can he or she be acting in breach of some Kantian deontological requirement. My high-carbon lifestyle cannot be represented as using others solely as means rather than as 'ends-in-themselves', the most compelling way of formulating a Kantian moral prohibition. I use the taxi driver solely as a means, in this sense, when I cause him to convey me across town and then decamp without paying my fare, because I make his aim of getting a living merely instrumental to mine of getting to my destination, instead of treating it as equally deserving of my consideration. (Again, we readily recognize the golden rule behind this way of thinking.) But since *my emissions* don't cause Australian bushfires, or the submerging of low-lying island states, or any other climate-driven adverse consequence, I can't be treating those suffering these inflictions as instrumental to my current lifestyle – if causality doesn't run one way, instrumentality can't run the other. (If it won't shift the nail, the rubber hammer can't be instrumental in knocking it in.) And I cannot use future people as mere means to my ends for the even simpler reason that nothing which doesn't (yet) exist can be instrumental to anything.

Nor, on the more abstract Kantian track, can my high emissions be wrong because I cannot universalize a maxim like "I will take a long-haul flight whenever I need to". Certainly, if everyone indeed acted on such a maxim, and nothing was done about the consequences, global overheating would rapidly become catastrophically final. But none of that brings my will into inescapable conflict with itself in anything like the way in which willing that everyone should adopt, for example, my maxim of lying when it suits me seems to do. There, having to universalize that maxim would appear to commit me to willing, at the same stroke as it were, both a situation in which I successfully deceive someone, and a situation in which no one could ever deceive anyone because no one would trust anything anyone said – and this does seem to involve a deep contradictoriness of intention. "What if everyone did that?", in other words, does not here point to any chance of this lie's seriously undermining the general practice of veracity, because if it meant *that*, "But they won't" would be a quite sufficient answer – rather, it flags up the golden-rule-shaped recognition that in lying I claim an exceptional status for my own case which the very *structure* of practical reasoning appears to prohibit. I can, however, quite happily will that everyone take a long-haul flight at need, as long as I remember also to will that something – who knows what … geoengineering the stratosphere, maybe? – be done to put the resulting situation right. Here, "What if everyone did that?" is straightforwardly answerable with "Well, in that still quite unlikely event this and that would happen, about which we could do such-and-such", and whatever the issues of contingent practicability thereby raised, this implies no internal will-conflict of the kind which is supposed to rule things out categorically on purely rational grounds.

In sum, the responsibility which is shirked by our failing to minimize our carbon emissions simply evades representation on any of these versions of a normative ethical template, whether applied collectively or individually.

Follow the footprints?

How, then, *can* we cash out convincingly the strong intuition which we nevertheless all have, that in contributing to global carbon emissions through forms of consumption reasonably describable as luxuries, I do something wrong? Perhaps the core of my moral responsibility here is to press for emissions-reducing action by an appropriately structured collective to which I myself belong – in my own case, by Great Britain – and I have a corresponding obligation of consistency as regards my own actions. Britain's emissions are undeniably a substantial contributor to the global total; its 'carbon footprint', 772 million tonnes of CO_2 equivalent in 2017 according to government calculations, seems extensive enough to be associated with significant levels of climate-driven damage. Moreover, it is still (just about) conceivable that Britain could be influential in this field by way of example and inspiration – its going carbon-neutral could give a real lead to the carbon-profligate global community, and its failing to do so is then evidently a failure to prevent, or at least try to prevent, preventable harm. So perhaps this is the same kind of case as when my country is engaged in practices, such as permitting slavery or waging an unjust war, which are clearly open to moral condemnation, and which its responsible citizens have a correspondingly moral duty to oppose. Nor, of course, could they do that without hypocrisy while still investing in war bonds themselves or owning even just the odd slave.

But in such cases, there are specific people – those being invaded or enslaved – to whom clear wrong is being done by a collective of which I am part; and the analogy to this situation superficially presented by the notion of an extensive carbon footprint to which I am contributing is really no analogy at all. That concept, and the way it is invoked in Thunberg's remark quoted a page or two back, actually doesn't bear examination. Considered more carefully, it reveals itself as essentially an artefact of the determination to represent responsibility here on the template of moral obligation. Thus it cannot without circularity be called in aid of that representation.

The carbon footprint of an individual or an activity, as standardly defined, is just the amount of greenhouse gas, in CO_2 equivalent, emitted by that person or activity. That is literally its carbon *contribution*, but the metaphor of a foot pressing down then smuggles in the wholly unwarranted additional assumption that this particular quantum of CO_2e will carry through to cause some specific damage in some specific area. Your actual footprint of course matches the size of your actual foot, and you are clearly responsible for damaging whatever gets trampled under it as you stump carelessly along: the larger your foot, the more things you squash down in the particular places where you tread, and if it is living things, and especially people, which are getting thus squashed, there is clearly a moral issue around. But, in the carbon case, things only get 'squashed underfoot' when billions of 'feet' are in play, and even then nothing is *squashed under* any particular one of those feet. To recapitulate in these terms the point already made about individual emissions, my foot on its own, even the 'foot' constituted by a whole lifetime's carbon emissions, doesn't just press down very lightly indeed, it leaves by itself no imprint whatever – it is always, as it were, treading on air. And this is also true of any collective (such as Britain) whose combined emissions could not by themselves alter the climate. In the nature of the case, no particular global-heating consequence such as an Australian bushfire is ever caused by British emissions, any more than the seawater used to extinguish it could be traced back to British rain. Only a collective large enough for its combined emissions to warm the world has a 'carbon footprint' in the sense which the metaphor tries illegitimately to generalize.

This metaphor in fact seems to be derived rather hastily from that of an *ecological* footprint, but the latter is a perfectly respectable notion, a comparison with which shows up clearly the illicit move involved in the former. An ecological footprint is an estimate in 'global hectares' (gha) of the biologically productive land and sea area needed to provide the renewable resources used, and absorb the wastes generated, by any particular set of activities. It is a variously useful notion which can be invoked to show (for instance) that in 2014 the global per capita ecological footprint was 2.8 gha, as against an available per capita biocapacity of 1.7 gha – giving us a handy single measure of the extent of global overconsumption vis-à-vis renewable resources at that point. The only comparable application of the carbon footprint concept is to provide a measure of how far some Western person's current emissions exceed what would be an equitable per capita allocation of the remaining global carbon budget, but here too it is only the total expenditure of this budget which causes the associated 'imprints'.

In fact causality really runs the other way in the ecological footprint case, and a better metaphor (though forgoing the effect of 'trampling') might be an ecological foot*hold*, or ecological *standing-room*. The fit of your foot to a particular land-area, as the latter term would make clearer, is part of the very idea of *resources*. Correspondingly, since any activity must be supported by some quantum of resource, there can be no ecological analogue to my carbon foot's treading on thin air. Everything I do, whether on my own or not, has an ecological footprint, but nothing done, even by very large numbers, really has a carbon footprint until we reach the level of the world community. It follows that there is no genuine analogy to be drawn between (say) a citizen's moral obligation to oppose the evil of an unjust war conducted by his or her country, and opposing the carbon habits of any collective whose behaviour the individual could reasonably hope to influence – since those in themselves, just as emissions, do not trample even potentially on anything or anyone by emitting CO_2. (Of course, the impedimenta of the carbon economy, oil pipelines and fracking for instance, may well trample on indigenous rights and all sorts of other things, but that is a different issue – and none of these infringements in itself jeopardizes life on Earth.)

The template of morality, in other words, is misapplied across the board to the climate crisis. But so what? Why should we be concerned by widespread reliance on it, or for instance by the quietly moralizing tone in exhortations to rectitude of which the following from the renewable electricity supplier Good Energy is a fair specimen: 'Tackling climate change isn't something that can be done on your own – we need everyone to play a part in reducing their impact on this planet. The best way to do this is to reduce your carbon footprint by taking your everyday carbon emissions into consideration'? Granted, this is confused (you can't tackle it on your own, and then suddenly you can), but should we worry much about that so long as it encourages people to review their consumerist lifestyles? A cheerful recent book even helps by telling you what the 'footprint' of each of a wide range of more and less ordinary activities, from sending a text via eating a banana to taking a trip on the space shuttle, actually comes to. After all, thinking in this way about their own behaviour is for most people a necessary first step towards supporting wider changes to the systems and institutions, national and international, which embed consumerism. Thinking about how to reduce your individual carbon *commitment* (which presupposes having some idea of what it is) is excellent preparation for coping with, and thus at a remove for acknowledging the need to sign up to, the very significant society-wide changes in relevant behaviours, including yours, which will have to be imposed sooner or later if catastrophe is to be averted.

More generally, why does misconstruing the climate emergency on the template of a moral issue matter, if it shakes people free of the prudential collective-action dilemma – the deterrent pressure to free-ride – and spurs or shames them into remedial action?

It matters because everyone knows, tacitly, that this is a kind of charade – more or less well meaning, depending on who is engaging in it. The widespread resort to moral language in which we don't really believe allows us to signal virtue – or, less self-regardingly, reassure ourselves that we are indeed acting rightly – without being irked by anything like the constraints that genuine morality imposes. So the person who finds that he or she just has to take a long-haul flight (maybe for business reasons, or as 'love-miles') is likely to take it without having to overcome anything remotely like the force of inhibition which would ordinarily confront a feeling that you 'just have to' kill some intolerable neighbour or set fire to their house. I can dutifully recycle, tot up my bananas and so forth, and think I am working in the right direction (and actually *be* so working, in the sense indicated two paragraphs back), but when I slip up I will experience at most the 'guilt' that goes with eating chocolate on a diet day, not that associated with hurting or betraying someone. At the collective level, political leaders some of whom (to give them due credit) would hesitate before starting an unjust war or reintroducing slavery, feel no corresponding compunction over repeatedly missing emissions or biodiversity targets to which they are notionally signed up, even while every credible expert is telling them that an anthropogenic mass extinction is in full swing. The template which everywhere spawns this kind of shadow-morality finds its most characteristic expression in the 'sustainability paradigm' which has crippled environmental action for thirty years and more, and which as we have seen consists in the solemn elaboration of standards, benchmarks, targets and rules, all purporting to register a moral imperative of concern for the future but all of which can (as everyone is, again, always tacitly aware) float and flex when the need for painful change becomes too pressing – their relation to genuine morality being fundamentally that of the commandment 'Thou shalt not steal except where unavoidable'.

Worst of all, none of this ever really persuades anyone but those already committed. The moral rhetoric for which climate campaigners instinctively reach tends to sound less and less compelling in the ears of the more and more broadly unconvinced, until it actually becomes counterproductive. Even many who might be inclined to pay attention to what the science says, or to what the media say it says, and to recognize a need for collective systemic change if only this could somehow take effect all at once, will remain firmly (and as I have argued, justifiably) convinced that they themselves do not behave *immorally* by going along with the system as things stand and while everyone else does, and they will increasingly resent its being suggested otherwise – never mind how they will react to the imposition of policies and regulations premised on straightening out their moral delinquency.

What is really, in the context of this book, the decisive point is that if, ignoring all the foregoing, we continue to force moral obligation into serving as the mode of engagement with climate crisis for which 'no one is too small', it will simply go on offering far too *pliable* a prop to support the weight of our hope against hope that catastrophe can still be averted.

Life-responsibility

That we do wrong by complicity in a way of living which is laying waste to life on Earth should nevertheless be starkly evident. The wrong we do springs from a much deeper source than morality, and the transcending of prudential self-interest here calls for a different kind of recognition from that which we are morally obliged to accord to the needs and concerns of other people. We have to grapple with and grasp this depth and difference in order to rely properly on the creativity of transformative hope.

What is at issue, in fact, is something very fundamental about how we understand human selfhood. To bring this into the open, suppose that for the reasons just explored I have no genuine moral responsibility or motivation either to refrain from high emissions or to campaign actively against them, and doing either would mean (as it clearly would) significant disruption or discomfort for me personally. Then it would appear that I am not self-interested in any morally suspect way – I am not being *selfish* – if, contemplating these options, I decline to take them up. For 'rational self-interest' is actually not just a tendentious economist's construal of human behaviour, but the representation in a mode apt to economic thought and procedure of what might be called the Enlightenment idea of the self as an autonomous subjective domain of perception, judgement and will, actively sovereign within its own private sphere. Its satisfaction, the pursuit of which must be correlative to any goal-directedness, is internal to it, as is its will: its dealings with objects in the world and other sovereign selves (its causal and ethical relations respectively, out of which its moral responsibilities are constructed) are external in the same sense. This way of understanding the self entails that my purposes, of any kind, have to make sense in the first place *as mine*. Fulfilling a moral duty makes important such sense, because of the nature of our interdependent involvement with others, but action which avoids harm or injustice to no one while imposing costs on myself, simply does not. (It fails the test of practical rationality as thoroughly as would pursuing some end which I really desire to avoid.) Thus, as long as we are operating as this sort of self, still overwhelmingly the default stance in Western culture, and are aware that no genuinely constraining moral obligations are in the frame, there should really be no evading the disincentive force of the collective-action dilemma as it inhibits individual climate-related action.

And yet – people do evade it. They speak out, and put their bodies on the line in civil resistance, and work hard at changing their consumption habits, and – more and more unmistakably as the crisis intensifies – they do so in response to an irresistible sense that they would be wrong to hold back on such actions as things now are. And like the small but indomitable Thunberg, they seem to be making a difference; or, if that is still a premature judgement, they at least seem to be generating enough impetus to provide us with grounds for hoping that a difference could yet be made – enough difference anyway to avert final catastrophe, if not the various disasters which are already inexorably on their way. So it would seem to follow that they cannot any longer be operating as Enlightenment selves.

What we are seeing in these recent, newly urgent manifestations of hope against hope, I suggest – and what this and the preceding chapter have put us in a position to recognize properly – is the emergence of a new kind of self, a new-created possibility for intelligible selfhood. Under unprecedented pressure, selves on the old Enlightenment pattern are being creatively reforged as entities with a different inner structure, a different relation between circumstances and purpose. There appears now, here and there, in answer to our wholly new situation and increasingly desperate need, the self no longer as autonomous rational agent – nor as throwback to the Christian or other religiously-configured soul – but as *centre of life-responsibility*. This is a self which takes with new seriousness its responsibilities towards the intergenerationality of human life and the continuing stability of the biosphere arising from its new circumstances and represented as insistences of life-hope in Chapter 1. But, crucially, it does not take up these responsibilities as moral duties – indeed, it does not *take up* such responsibilities at all, because they are internal to its being, so that there can arise for it no question as to the individual rationality of its embracing them. This means that we must think of it as a self which, essentially to its being and thus to all its purposes, belongs and answers to something other than itself within itself – something which it makes sense not only to call *life* but to see as resonating with the wider life now under threat.

How could something other than myself be within myself in the relevant sense? – not, of course, the sense in which my kidneys or gall bladder are 'within myself', but more like that in which my will is internal to my being. And how can something which holds me responsible be as essentially internal to me as my will and also pervade the biosphere? In order to remain undaunted by such questions it is worth stressing that we are not talking here about anything remote from ordinary experience. We are trying to explicate the refusal to be part of a climate- and life-hazarding system which we are seeing now with increasing frequency, especially among the young. When someone judges that in the interests of sustainability as properly conceived, there should be (for instance) a law limiting recreational flying, and determines both to push for such a law and to make personal travel choices as if there already was one, *however little measurable difference that might make in either case*, he or she is on the way to self-reconstituting as a centre of life-responsibility as I intend the term. Again, this is really no more than to act in the spirit of E.F. Schumacher's old advice to do the right thing and not bother your head over whether you are going to be successful, because unless you do that you will be doing the wrong thing. But the right and wrong things have to be understood not as morally, but as *vitally* right or wrong, and the kind of *integrity* at stake is not to be explicated in moral terms (as, perhaps, steadfastness in placing oneself within the jurisdiction of a law which one is prescribing to others), but as a kind of vital internal coherence.

To what, in pursuing such coherence, is one responsible? I seem to have full sovereign authority over my own intending – which is more than just a matter of wanting to do something and believing that I shall, but involves what we might describe as deliberately aiming my loaded will in a particular direction. Such intentions are of course responsive to changes in my relevant desires and beliefs – I can intend to go running, but if I stop wanting to (maybe I'm just too tired), or if my belief that I can find my running shoes turns out to be false, then that intention will come up for revision. But given my desires and beliefs, it goes with the model of self as private autonomous domain that the irreducible pointing of my will which takes me beyond them towards action is not answerable to anything else internal to myself. All its answerability on this model is moral, a matter of constraints arising from relations with other selves, or prudential, a matter of their arising from relations between my own present and future states.

But, as we are all intuitively aware, there are important kinds of occasion when this isn't actually true. I am thinking here of the sort of situation in which Lawrence's injunction to find and follow your deepest impulse is relevant; occasions, for example, where your conscious self forms an intention against which there is a deep inner protest on behalf of something in you which isn't at the disposal of autonomous will – or equally, where your conscious self is quite uncertain what intention to form, and listening for an inner voice from beyond the will becomes decisively important. "All my reasons and duties were driving me to do that, but I also knew deep inside that it wasn't right for me" – one can feel this about a job choice, a crucial step in a relationship, commitment to a defining intellectual position, a prospective marriage, indeed any life-significant action from which there can be no easy turning back. In all such cases, fully ('wholeheartedly') to intend the action requires that my consciously directed will has to coincide with, or at least not override, this kind of inner prompting or resistance. And what the conscious will has to coincide with here is something internal to me but distinct from me, such that I can no more simply cancel it out by an exertion of that will than I can simply override your non-compliance in declaring some joint intention of ours to act.

That there is this pressing inner complexity of the desiring self when intentions are formed in such cases is not denied, though it can be obscured, if we are inclined to say that consulting my deepest impulse is just how I find out what at bottom I *really* want. For then within that supposedly unifying category of *desire* we should still have to distinguish two strikingly different modes of conative force. There are on the one hand the kind of desires which report, as it were, to the conscious ego, and are mine to manage: those from which I can step back in critique and evaluation, asking from the security of my central, authoritative self whether and how far they give me reason to act. Such desires, we might say, are answerable to me. But there must also be recognized, on the other hand, the deep desires, ignored at one's peril, to which *I am answerable* – those which are not so much a case of 'what *I* really want' as of recognizing something wanted as and through me. These latter are such that if I fail to act on them I am failing vitally. This is the sense it makes to call what I am answerable to here, *life* – a principle of harmonious development, or *nisus* towards flourishing, of my whole being, that whole which as Schopenhauer noted is necessarily dark to subjective consciousness. And this, while evidently internal to me, is also distinct from me in not *belonging to* the subjective self which is the conscious bearer of rights-claims. It is something essential to me in the way that my individuality is essential, but not unique to me in that kind of way; it is something larger than me in which I *am* me by sharing.

So far this notion of inner answerability might be held to recognize the self as responsible only to what is beyond the will, but as it were privately, within itself. But now, because our contemporary situation is such that ecological knowledge of human dependency on the biosphere has become so much more readily available than before, and with it an awareness that humankind as a whole is now so drastically travestying that dependency, some selves are beginning to generalize this intuitive sense of *living* responsibility, and move from alertness to life as an impersonal manifestation of energy in one's own case, to sensing the resonance of that energy with a wider principle of flourishing across the biosphere, of which one's own case is one local, first-personal expression. Such affirmation must mean accepting that one's own deepest life-impulse, when one listens for it in forming significant intentions and then acting answerably, now resonates also with the voice of threatened life at large.

This I think is the recognition really underlying the idea of Self-realization in Deep Ecology, where the individual self is supposed to *identify* itself with other forms of life, all the way up to the biosphere, considered as nested self-realizing systems within an all-embracing holistic unity. As Freya Mathews puts this, glossing the late Norwegian philosopher Arne Næss:

> When we recognise the involvement of wider wholes in our identity, an expansion in the scope of our identity and hence of our self-love occurs [leading to] a loving and protective attitude to the world – an extension of our loving and protective attitude towards our own bodies.

But this way of framing life-responsibility gives far too many hostages to incredulity for present purposes. It is not just that, pushed to specifics, it threatens absurdities (no one is ever going to recoil from the logging of the rainforest in anything remotely like the way she recoils from taking the secateurs to her toes), but that the cloudy rhetoric of holism points us in exactly the wrong direction for the kind of agency which we now need. Loving or caring, especially in challenging situations, must involve attention to the interests of *another* – love, as Iris Murdoch wisely remarked, is 'the extremely difficult realization that something other than oneself is real' – but expanded self-identification in a spirit of 'All is One' tends to dissolve this basis for such exacting attitudes, whether directed towards other people or aspects of the natural world. Understood as a centre of life-responsibility, however, the self remains active bearer of perception and will, sharply distinct from other agents and objects – an *individual* source of the creative energy which only arises in individuals. But the quality of its individual action is changed, since what it has to answer to, the going of its conscious will with the grain of its own deepest life, now resonates increasingly to the jeopardy of the wider life which that depth transmits and expresses.

Creative power and social transformation

These are suggestions and indications which need completing (insofar as they can be completed) by the whole book. But that potential power must be the concluding note here, and it needs to be reconnected with the idea of creativity in order to emphasize the unity of this and the preceding chapter.

It is crucial to insist that the self as centre of life-responsibility must indeed be recognized, wherever it is found, for something *newly formed*. It is certainly not the old Enlightenment sovereign subjectivity with its moral duties extended to new kinds of entity, as in standard environmental ethics, nor with the scope of its self-awareness and self-concern broadened indefinitely, as for Deep Ecology – however plausibly or implausibly in either case. Rather, it is a conscious personal focus of an impersonal force which is only 'there' to inform action at each individual focus, but which *is* always there and available to act. Hoping against hope in the face of our unprecedented crisis has to mean hoping that one's agency can somehow make a difference against all the odds, and the kind of self of which that could be true – the self with that inner relation to responsibility, undaunted in its exemplary action by any hesitation over the rationality of its likely impact – is forged in the very same stroke, because to trust oneself to such active hoping is to release the life-energy which creates its possibility.

But this brings us up against a crucial question which many readers will for some time have been wanting to ask. How does recognizing that capacity to create new possibility at the level of individual action translate into seeing as realistic the kind of very widespread *social* transformation on which avoidance of climate and environmental catastrophe now depends?

There is no easy answer to be found here by thinking of individual life-responsible transformations in attitude and behaviour as accumulating until they become sufficiently numerous to trigger a 'tipping point' moment at which society-wide transformations will result. The trouble with this is not just the long-familiar bind that until key social changes have happened (say, in the field of transport policy and provision) many of the individual transformations supposed to drive them cumulatively are effectively precluded. A much deeper difficulty is that the whole 'social statics' model in which actions taken by individuals 'snowball' into society-wide change, remains wide open to the collective-action dilemma. People can no doubt be expected to change their behaviour as they are influenced by seeing others (such as Ms Thunberg) doing so in an inspiring way. But they will only change as far as it makes sense for them to do so within the framework of general conditioning factors – lifestyle, habit, sunk costs of existing systems and practices, media-fostered comparators and expectations, the legal and social framework ... – all of which will continue to operate very powerfully. The dawning of life-responsibility means, as we have seen, a radical change in the internal relations of the self to its ecologically relevant actions. But as long as the relations of selves to one another are still conceived as *external* relations, and influencing therefore as essentially a psychological form of pressure, we have no reason to expect that defection will not continue to dominate cooperation even in the matter of response to such influencing. People may indeed fly a bit less when they see prominent others doing so, or when alerted to the issues by some eye- and conscience-catching demonstration; they may, similarly, eat rather less meat, do rather more recycling, cut down to some extent on their energy usage – or, they may 'whole-hog' in one or even a couple of these areas (completely revise their diet, for instance). But the weight of the whole disincentive structure will ensure that these changes go only so far – with full life-responsible commitment withheld until everyone else is seen to be so committed, and thus never arriving. The snowballing effect, that is, will be constrained by the possibilities already inherent in the starting situation: the ball will roll as far and as fast as the forces of resistance will let it – and these forces are both very strong and firmly in place before it even begins to move, which brings out the real nature of the conceptual challenge here. For creative power to transform possibility in defiance of such empirical likelihoods belongs, as we have seen, to individuals: it is only as bearers of a will capable of hoping against hope that they can turn themselves into centres of life-responsibility. But there is no analogue to individual will at the level of social collectives if we conceive of these as accumulations of separate individuals, the possibilities for which can therefore only be determined by general laws of interaction like those of any other aggregated mass.

Here, however, it is vital to identify and resist a misunderstanding of the term 'social' which goes with the Scientific World View, or rather, with its inappropriate (scient*istic*) extension beyond its field of useful application. We need to recognize that societies, while indeed not super-individual entities with subjectivity and creative freedom of their own, are nevertheless not merely aggregates of individual human beings with those attributes. Those alternatives are not exhaustive. With this we reach something absolutely fundamental, which therefore cannot really be *explained*, only indicated – but unless we grasp it, the whole argument of this book will misfire.

The nature of the third alternative emerges perhaps least mistakably if we think about the mode of existence of language. This transcends the minds of individual speakers, but not as something *spatially* beyond them (as are, for instance, the dictionaries in which lexicographers, or the bytes in which corpus linguists, try to capture it). We can't make clear how it does exist and how we share it, in any other way than by pointing to facts with which everyone is intimately familiar, but with a kind of familiarity which it is hard to bring into the light of full, explicit recognition. Among these facts, for instance, are that we are not sharers in a language in anything like the way we could share membership of some group – you with your individual membership and me with mine in parallel. We get a sense of the relevant kind of sharing when we note that *in language*, I can share your thoughts without being (impossibly) inside your head, but without your thoughts being anywhere *outside* your head. (The marks or sounds on which I rely, and which you will have produced in response, say, to my asking you to "Give me your thoughts …", are indeed outside your head, but aren't themselves *your thoughts*.) Again, when I speak and mean what I say, I must do so in words which have meaning only because others have meant the same thing by them. But 'the same thing' can't be just what *I* am meaning here, which is a matter of my intentions springing out of my particular situation; and nor do words themselves, aside from use, mean anything – again, in themselves they are just marks or sounds. So I would seem precluded either way from choosing them to embody *my* meaning – but equally clearly, that is exactly what I do.

In face of such facts, we can only fall back on saying that a language constitutes a public or trans-personal realm, radically disanalogous to physical space, in which different minds can *meet*. Here again, as with society as conscious super-individual entity *or* statistical aggregate, we have a duality – inner mental arena *or* outward physical space – which looks as if it must be exhaustive, but which we can't help recognizing not to be. And once we see that we have to acknowledge a domain of collaborative-creative intersubjectivity in order to make sense of the way we share our meanings in language, we can say that the human collectivities which our linguistic powers play such a central role in bringing into being subsist in that sort of space too, and have that kind of reality – the reality of the *human world*. In this light we can see, without at all impugning the unique individuality of individuals, that the concerns and value-recognitions which prompt them towards life-responsibility are *social all the way down* – as meanings, they are radically transpersonal in their very essence – and the corollary of that is that individual creation of new possibilities for human being is transformative all the way up.

Admitting these truths, it should be noted, calls for the same order of recognition as does our ability freely to create configurations of possibility which did not precede or condition our choices – it demands the acknowledgement as irresistibly real of something for which the objectivist-scientistic picture of the world has no conceptual room, and which depends on taking our irreducibly subjective living agency with fundamental seriousness. Just as my unhindered choices are made from a position of being *actively inside* my own conditions, and thus do not bear causally *on* me as they do on the objects which they affect, so we each understand the significances in which we share with those whom we meet in meaning *individually from inside*, making that collaborative encounter radically unlike any kind of collision of forces external to one another.

'Social transformation', it then follows, is not something which happens when enough individuals accumulate, on the model of a momentum building up to overcome a resistance. Rather, it consists in significantly new meanings and values' being put into general currency – something which can only be initiated by individuals, but which is *already* socially transformative when the creatively acting individual takes that initiative. The 'tipping point', if we continue to find that metaphor helpful, arrives when new understandings and values come to make irresistible shared sense across the meaning-space of a relevant community: the proper analogue here, however, is not any shifting balance of calculable forces and resistances, but the way in which the expressive possibilities of a literary or artistic form are sometimes found to have been decisively altered by the contribution of an innovator of genius. And when meanings which are shared in this *sui generis* way do start to alter unignorably, all the 'social' institutions, critically including laws and structures of governance as well as habits, expectations and taken-for-granted assumptions, will start to follow suit – in what will always be as painful and disruptive a process as anything else creative.

It remains true, of course, that none of this can happen without inspiring example and its increasing imitation, expanding the participation of (in particular) thoughtful and articulate people from all walks of life in events and occasions which are expressively constitutive of emergent meaning-change.

There remains also the very urgent question whether in our current plight this can happen anything like fast enough for the necessary transformations to occur. But properly recognizing the sociality of these processes emphasizes how they are seamlessly continuous from the free creative action through which individually unique human beings are enabled, alone in nature, to be the sources of possibility and not just one of its manifestations.

It should emphasize, too, the absolutely vital role of charismatic leadership in these transformations. The essence of such leadership is not only to shape by example an emerging common practice, but also to articulate for the relevant group the shared meanings which, unled, they would be unable to formulate. This articulation shares with both art and free action the nature of expressive-creative activity – it brings into being something real which may not have been among the relevant possibilities beforehand. Life-responsible leaders of that order will be needed at every level, from local community through national government to supranational institutions, if individual transformation is to set larger and larger changes in motion. In getting "We *create* possibility" to become the acknowledged master-principle of public action – where *we* are social beings in the sense of 'social' just elucidated – the role of such leadership will be utterly crucial. Whether we shall have the epochal luck actually to find leaders of that calibre and courage in our increasingly desperate straits, remains to be seen. The range of considerations addressed in the concluding chapters of this book bear on the conditions within which they will have to operate, and if it helps to arm even one or two of them with the concepts to support them in their task, the book will have served a major part of its purpose.

All this might also suggest the paradoxically encouraging thought that things had to get as bad as they have got for us to have gained any real chance of making them better. The collective-action dilemma explains in terms of the dominant Western self-conception why, in comparison with the scale and seriousness of the climate and ecological crisis, so very little has happened over the past forty years. The prudential irrationality, on those terms, of individual or group self-restraint has manifested itself as the huge, system-wide inertial drag on motivation towards change, the resistance which has confined concerned groups to minority status, Green parties still mostly to the political fringes, and what institutional change there has been to the essentially tokenistic – the adoption of 'sustainability' constraints which never, when push comes to shove, actually constrain. Of course there have been other factors in play, such as corporate greed and malfeasance and the abysmal quality of telly-democratic politics and politicians, but this systemic inertia has been key – for corporate deceit and political short-termism, however culpable, have largely been in the service of huge public reluctance to contemplate any significant interference with fossil-fuel lifestyles. So there could have been no prospect of breaking out of this impasse without a reshaping in (initially) one or two creative individuals of that dominant Enlightenment sense of the private, autonomous self. And that in turn could not have happened before we had reached the present point, where empirically based hope has visibly begun to fail any test of realism, and hope against hope has emerged as all we have left. For the bearers of such hope, unable to reassure themselves with probabilities and recognizing in their guts that morality is far too frail a prop, have had to *reach down to life* – to the instinctive drive of life-hope in the individual which now so powerfully resonates with the need to defend life at large.

In doing so, they reach for something represented by the magnificent bay stallion St Mawr in Lawrence's tale of that name:

> He stands where one can't get at him. And he burns with life. And where does his life come from, to him? That's the mystery. That great burning life in him, which is never dead. ... But think ... if we could get our lives straight from the source, as the animals do, and still be ourselves ...

Getting our lives from the source is life-responsibility. Our dreadful mechanical-egalitarian civilization, with its draining irreverence, its willed perversities and its attempted domination of the wild in nature and in ourselves, had seemed already in 1925, when Lawrence wrote that, to have hugely narrowed humanity's chances of recovering such openness to the central fire. But now, in the final destructive self-undermining of its own premises, it demands from those who are paying attention nothing less than fighting tooth and nail for recovery if we are to survive at all. For to pay proper attention must be to recoil in appalled horror from what humanity has since then made of the whole Earth – the mess, the depletion, the barbarities of concrete and plastic, the barrenness where there had been living richness, the endless human multiplicity, the grubby fingerprints of our greed in the oceans and on the very atmosphere – and positively, to embrace youthful energy and hope against hope as our only resources for regaining life-balance.

'This civilisation is finished', to quote Rupert Read, one of the very few honest writers about these issues; and for all the pain that its ending will involve, it is good that we are forced to live dangerously in seeking out, at last, something newly vital to carry us through to its successor – if there is to be one. And we must believe that the search, directed by that kind of inner answerability, can release hitherto untapped forces. 'No one is too small to make a difference' is not in this light a conclusion from wide experience, nor a putative law of human behaviour vulnerable to sardonic empirical comparisons, but a brave affirmation of life-hope – so compelling, just because it so clearly comes straight from the heart of an evidently still *in*experienced young person. (*Of course* someone that age will be inexperienced, and that is no put-down but a recognition of what is essential to the raw power of her message.) This gives us, too, the general structure of the hope invested in transformation on which I think all climate activists are really now relying, even when they are driven by dominant cultural habit to invoke unpersuasive precedents in order to keep up their spirits, or to express their energy in conventional terms of moral outrage. "It's not too late, because if you think it's too late, where's the drive to act immediately?" is indeed chocolate-cake logic from the perspective of empirical realism. But we can perhaps now fully appreciate what it says in the light of transformative realism: that the life-energy which manifests itself in going on hoping, against all the odds, that it's not too late, is what empowers us to create – literally, to bring into existence as real – the possibility of enough of us becoming the kind of people for whom it *may* be not (quite) too late. It is ultimately the creativity of life-hope which may, just, turn things around.

5

The Bounds of Utopia

So the realism of transformation is ambitious. If hope informed by it can envisage a rapid and dramatic shift in the perception by individuals of their agency and motivations, in relation to their whole Earth-systemic context, that shift could also by extension transform the pressures shaping action by groups and collectives all the way up to the nation state and the international order. This means we are to hope for nothing less than a new kind of movement for change, establishing itself with astonishing speed through all the new forms of connectivity now available – a movement of deliberate and emphatic individual acceptance of responsibility for the wider biospheric life which is now threatened. How such a movement might be brought sufficiently swiftly into being, through what activities of consciousness-raising and mobilization coupled with the impacts of which unignorably climate-driven disasters, is an open-ended matter. So are the expressive forms which it might take, and the drastic political changes which it will demand. Empirically, none of that is remotely credible. Practically, we have no option left but to hope against hope that it could yet happen. That means believing what the previous chapters have tried to set out the warrant for believing: that the hope which we must invest counter-empirically in bringing transformation about can genuinely create the possibility of our becoming the life-responsible agents of its happening.

Counter-empirical hope, however, remains *hope*: desire for a valued outcome under the sign of contingency. As such – the point from which our whole enquiry into its implications started – it must be directed at something which, while uncertain of achievement, is nevertheless in principle achievable, which means, something *really* (at some level) in prospect. That is, it must still, for all the life-depth of its impulsion, meet the condition of realism by pointing to real possibilities. But now, if we have recognized the power of human action spurred by hope to *create* possibility, what kind of requirement could that be? Haven't we presented ourselves with an all-purpose superpower to do anything whatsoever? The suspicion that we might be trying to do just that lurks, naturally enough, not far beneath our ongoing cultural reluctance to acknowledge what I have been calling the creativity of human action. But, somewhere between the aspiration towards that kind of world-changing movement and the aspiration 'to put everything right' – the high note of uplift on which the book co-authored by Greta Thunberg and her family very understandably ends – must fall the line to be drawn between hope, even against hope, and wishful thinking. And what sense, once we have set empirical realism firmly aside, could it make to lay claim to an open-ended human capacity to bring forth the decisively new and unforeseeable, and yet to suppose that we can still reliably draw that line?

Why should being able to draw it matter, as long as we remain undaunted in practice by the odds which we are facing? To answer that question, we need to ask another: what happens when the mobilization falters, as it surely will – when the authorities start seriously to lose patience with disruptive direct action, for instance, or when key charismatic leaders burn out and the collective energy of which they are generative sources wanes? What strength have we for remaining committed when the supposed precedents for transformation are seen to have been too glibly invoked? What happens when, correspondingly, the precedents for *failure* begin to loom large? (And global communism, post-war internationalism, gender equality and the liberating power of the internet are all examples of hugely ambitious promises which have not been delivered on). What happens when "Yes we can" turns into "Maybe after all we can't"? (That form of the question, at least, already has its dismaying answer from across the Atlantic: what happens is Trump; and what would we do if not just one sick late-capitalist society, but the whole movement of life-energy to rescue the human future, were to find itself Trumped?)

Campaigners have many *ad hoc* resources of mutual support and discipline for when they have to keep going through bleak patches. But this is an existential crisis with the whole human future on the line, and more seems called for. To arm ourselves in advance against the especially bitter kind of disillusion that must come with any stalling or apparent failure of the deliberately courageous rejection of disillusion in such a crisis, we need our commitment to transformative action to depend on a strong life-confidence in human creativity. And that in turn will depend on a clear understanding that our hopeful energies are not, after all, being poured away in wishful thinking which is ultimately going nowhere.

Transformative realism and reality

How do we build and support a realistic understanding of human creative power – of how far the possibilities which hope can generate extend *in reality*? One approach, naturally appealing to a philosopher, would require argument which could not be scientific, social-scientific or historical. At issue would be the question of how the creation of radically new possibilities through human action could form part of the underlying structure of things in the way to which I was pointing in Chapter 4. That is not something on which any form of empirical explanation could be the final authority, because empirical explanation only works through the assumption that universal causality constitutes the world's underlying structure, and that assumption itself would be in question here. Instead, we should be entering on the terrain demarcated by the German-American political philosopher Hannah Arendt when she wrote that:

> With the creation of man, the principle of beginning came into the world. ... The fact that man is capable of action means that the unexpected can be expected of him, that he is able to perform what is infinitely improbable. And this ... is possible only because each man is unique, so that with each birth something uniquely new comes into the world.

And whether or not we endorse Arendt's particular conceptual system, it is evident that something like this would represent the *kind* of argument which would be called for to establish in any theoretical fashion the realism of transformative hope. That is, it would be an argument appealing to the fundamental nature of human being considered not anthropologically or ethologically, but from an existential-philosophical standpoint.

That need not be an unduly scary prospect for the non-philosophical reader. It depends on the quite accessible general point that 'what experience tells us' is always going to be a matter of what interpretive framing we set up for it. For instance, to take an example from Al Gore's more recent film: when we are caught up in the transformative energies of Gandhian peaceful protest, do we directly feel *satyagraha*, 'the force of truth' (or, the creative power of life-hope, if we prefer that idiom), moving through us? – or do we have a psycho-sociologically explicable sense of exhilarating group solidarity which we instinctively mythologize in those ways? Evidently, no appeal to the experience just as such is going to answer that kind of question, since it is the ambiguity of our felt experience as between those two interpretations which gives rise to the question that we are asking. Again, is our undeniable experience of deciding freely to act an indication that we do radically initiate changed states of the world and realize new possibilities in a way unique to human beings, or is it just the way we feel when no one is preventing our doing whatever we are caused to want? – again, no appeal to our experience *as of* free agency, which experience is not in dispute, could answer that question. So an account like Arendt's has to be doing something else. And what it is doing is arguing, in terms of an old philosophical distinction, *a priori* rather than *a posteriori* – it is exploring the framing concepts which as such precede rather than depend on what we take experience to show us. In pursuing such an argument, we don't completely prescind from experience, which would leave our thought stranded, but nor do we appeal to it to support our conclusions – rather, we hold it in mind with an eye to which framing seems to make the richest sense of it.

But this means that argument about the scope of our creativity would have to start from very general considerations about how human beings *must be*, for their experience of creative action to have the basic character it has, and it would have to proceed from that starting point on conceptual grounds. Argument of this nature will inevitably be 'abstract', and many people concerned about hope and the climate emergency may not be much interested in that mode of thought. There is also the methodological worry that no argument of an abstract kind looks as if it could *ground* life-confidence in the power of human creativity, because if we lacked such confidence, how could mere argument persuade us into it? (Essentially the same point is made in relation to religious faith by Paul Tillich: 'Faith precedes all attempts to derive it from something else, because these attempts are themselves based on faith.') These considerations strongly suggest that we should find some other way to shore up confidence in transformative thinking as realistic. Here, I believe that exploring the issues raised by utopianism can help. Having done so, we may find, as before, that the metaphysical ground has started to shift decisively under our feet.

Utopianism and responsibility

How can we believe in hope as open-endedly capable of creating new possibilities, and yet recognize and guard against any tendency inherent in it to mislead us as to how far that capability extends? The best way to get a practical handle on this question is to ask: how can we be responsibly utopian?

The word *utopia* is ambiguous. Etymologically, it renders Ancient Greek *ou-topia*, meaning literally no-place, or by extension, a non-existent place. Explicit utopias from Thomas More's onwards have always been imaginary localities. But they have been so, because in them life has been pictured as going very much better for their inhabitants than in the actual world contemporary with their imagining – hence the parallel resonance of *eu-topia*, a good place. Hence too the positive sense of the term *utopian*, in which it means presenting us with a very significantly better human situation than our own, one which doesn't exist but could (perhaps) be brought into being if we summoned the necessary commitment and goodwill. In this sense, life-hope invested in transformation may fairly be called utopian, and it is by courageously embarking ourselves in such hopes that we create possibilities for their realization. Everyone remembers Oscar Wilde in this connection: 'A map of the world that does not include Utopia is not worth even glancing at, for it leaves out the country at which Humanity is always landing.' As Krishan Kumar glosses this, 'Utopian conceptions are indispensable ... without them politics is a soulless void, a mere instrumentality without purpose or vision'. The fact that utopianism has always been a very characteristically human mode of thought (starting with Hesiod's Golden Age, and surviving the non-delivery of a long list of specific utopias since) is itself an indicator of the life-depth of hope.

But then there is also, of course, what Dr Johnson called 'the vanity of human wishes'. An ideal society may be non-existent just because it happens not (or, not yet) to exist, but it may also not exist because it *couldn't* exist – because in imagining it, we have gone beyond the kind or degree of betterment possible for human beings, however transformed (and doing this must always be a danger as long as we are not, after all, equipped with magic powers). Hence arises the negative meaning of *utopian*, as describing someone who fails to recognize that his or her transformative aspirations are literally out of this world. While the pragmatist or disillusioned realist takes human beings as they more-or-less ordinarily are, and the idealist wants them to be the best they can be, the utopian in this negative sense wants them to be, impossibly, something even better.

Utopian thinking, however, plainly doesn't set out to want more than is humanly possible. Whether any particular hope is just too idealist for implementation in 'the crooked timber of humanity' isn't always apparent beforehand – if it were, the idea of transformation would find no place. Utopian thinking is so indispensable politically and practically just because we often can't tell in advance whether or not what we are hoping for presumes too far on the possibilities which we might bring into being. That is why proposed utopias can serve to keep us aspirationally up to the mark, to ensure that at any rate we never set our sights on less than the best, nor take for granted beforehand that we know how much of the best is achievable. And this is indeed the role which explicitly Utopian writing has played, from More through Morris (and also Marx, whatever he himself thought he was doing), all the way to Macy and other contemporary practitioners. A natural first characterization of *responsible* utopianism, then, would be that it performs this aspirational role while retaining the flexibility to correct continuously for the difference between the genuinely achievable and what goes beyond it, as we discover ongoingly which is likely to be which.

But of course, this only works insofar as we are ready to rely on our ongoing experience to demonstrate which *is* which. The serious trouble in this connection with the counter-empirical domain of hoping against hope is that there seems to be no way for this crucial distinction reliably to emerge – because ongoing self-correction in response to the contingently emergent is just what hope invested in transformative possibility seems to rule out. Insisting that transformation always remains possible whatever the precedents and the history to date, it precisely removes the check constituted by finding out what may really be possible from what does actually happen.

Is it responsibly utopian, for instance, to aspire to 'a world without weapons', something which Macy and Johnstone celebrate as one goal of their 'Active Hope'? Firmly renouncing (as they claim) wishful thinking, they cite as evidence of pragmatic intent a Quaker workshop deliberately structured around fostering these aspirations:

> To help participants get into … fantasying mode, the second exercise is to step into a childhood memory. … After a couple of minutes of remembering, each participant turns to a neighbour. … Sharing the experience helps clarify the memory. It is this type of imagining you do when you image the world thirty years from now. You will see a world without weapons around you with the same vividness.

Now clearly, while such supportive interactivity could well develop a strong sense of shared commitment and common purpose, group self-delusion is still delusion – and arguably more dangerous than the merely individual kind, because of the momentum of reinforcement which groups so readily develop. But when they think of themselves as embracing creative hope, even world-disarmers bolstering their pooled fantasies can turn the accusation of irresponsibility aside: they can perfectly legitimately say: "Of course, we've seen no such transformative shift in this direction so far: but ... ."

While avowedly utopian thinking brings this issue into relief most sharply, the difficulty is a completely general one. The counter-empirical hope on which we must now rely affirms that *whatever* may have happened up to now, and whatever future our past experience may now make highly probable, we can't know that there won't be a transformative leap to a new set of circumstances. The corollary of that, however, is that we equally can't know that there will be. In the climate case, it is after all quite plausible to suppose that the adverse forces driving business as usual – the corporate conspirators and their political cronies, the technological and other systemic imperatives, the firm attachment of Anderson's 'few per cent' to their (that is, our) unprecedentedly comfortable lifestyles, worldwide aspirations to lifestyles of that same order – all these might well in combination be powerful enough to prevent it. But explicitly counter-empirical hope can then turn into a way of insuring itself against taking this outcome seriously; it will have immunized itself against the mere failure of hoped-for changes to have materialized by any given point, since we can always reassure ourselves that all we have encountered up to that point is failure to have reached the transformative tipping point *yet*. And if we are relying on a shift the arrival of which is wholly unpredictable, then its not having arrived by any given point will of course be just what we should have expected, and no cause for revising our expectations. The merely contingent absence of transformative change can always appear as only ever a temporary postponement.

This might seem to explain a lot about the recent history of climate change activism. If preventing disastrously runaway climate destabilization can now only be the object of counter-empirical hope, still that means we need never be knocked back by the failure of any of a succession of necessary steps in this direction – Conferences of the Parties, emissions reductions by set target dates, plans to roll out aspirational carbon-capture technologies and so on. As each of these hoped-for recourses fail to materialize or to take remotely adequate effect (which has been the constant pattern from Rio 1992 onwards), we simply reassert our commitment to pressing on until the unpredictable transformative shift, kept in play by the very fact of our pressing on, actually arrives. Hence the apparently inexhaustible capacity of such commitment to survive each 'last-chance-to-stop-dangerous-climate-change' along the depressing trail from Kyoto via Copenhagen and Cancun to Paris and onwards. Nothing counts as failure because, if hope against hope can create possibility, transformation always remains possible, so that this process need never run out of road. In previous writing I have identified such entrenched activist persistence as a form of denial, but we can now see why this doesn't quite meet the case. Denial refuses to recognize looming defeat for what it is; but this kind of utopianism looks as if it can do even better than that by setting itself up so that *nothing will ever be able to count* as defeat.

Hope and defeasibility

Evidently what is threatening to go wrong here is that utopian aspiration begins to lose touch with reality when the hope invested in it is not *defeasible* – when the counter-empirical hoping which tends to support it in the most demanding cases cannot, of its nature, be defeated by disappointing experience. But defeasibility is as much a condition of genuine hope as is realism – it is, in fact, merely the other face of the condition of realism.

That condition, as we have seen, requires that an outcome has to be something with a real chance of happening, for my attitude towards it to qualify as hope. (You can test this by trying to hope that two plus two equal five from next Tuesday, or that Attlee return as Prime Minister – your failure to get as far as recognizably hoping will not be because you haven't tried hard enough, however hard you try.) But the corollary of there being a real chance that something will happen is that there also has to be a real chance of its not happening. It is not just that hoping often exposes us to disappointment, but that in a situation where the outcome is genuinely open, we must hazard disappointment to be *hoping* at all. Hope, even at the most mundane level, is an embarking of oneself in chance. (Again, try hoping that gravity will keep on working – this will feel a bit like trying to leap towards yourself from where you are now standing.) Hope always requires us to *risk* something, from momentary disappointment all the way to kinds of lasting desolation which can unhinge us – hence its close association with the virtue of courage. And this stance towards the future seems to go naturally with life; the old Stoic injunction to keep oneself safe from disappointment by keeping one's hopes minimal is ultimately a counsel of futility. That means we have to live in constant self-commitment to contingencies, from the trivial ("I hope it doesn't rain") to the variously serious ("I hope we can find a way to stay together", "I hope we can repair the NHS" ...), all the way to "I hope climate-driven catastrophe can be averted, however terrible the prospects look". And then that hope, which looked like commitment at the most demanding end of the spectrum where it becomes something like a declaration of faith (and one on which we are staking the human future), suddenly threatens to risk *nothing*, because its very nature is to be immune from contingency.

This point mustn't be misunderstood. Activists inspired by hope against hope often risk a very great deal, from personal inconvenience through distress and burnout to fine, imprisonment and even (though not, or not yet, in Britain) violent death. At issue here, however, is risk of a different order: that which is precisely *not* run by someone whose credo and practice are so framed as not to expose them to the chance of finding themselves fundamentally mistaken. And immunity to that risk not only impugns hope but threatens honesty. If we are deliberately closed, as counter-empirically we are drawn to be, against emergent evidence that what we are hoping for lies beyond what is actually achievable, the line between this kind of hope and playing false with our perception of reality becomes very thin indeed; and sooner or later we will step across it, from going on hoping for an outcome because we refuse to concede that it isn't any longer possible, to refusing to concede because our commitments, or our moral comfort or our mental equilibrium, depend on hanging onto hope. And that, of course, is to have crossed the line from hoping into wishful thinking.

So if there is to be a criterion by which we can stay realistic here, it looks as if it will turn on whether counter-empirical hope can be entertained in some sense defeasibly. For if the possibility of failure necessarily shadows any real possibility of success, it must so shadow even the possibility of success which our active hoping creates. Thus while we cannot both believe confidently in that creative power and appeal to the achievability of what we are hoping for as a reliable test of our realism, we can nevertheless always take its lack of defeasibility as a mark of *un*realism. But then, if such hope is inherently immune from defeat by ongoing experience, we need to identify a different form or mode of defeasibility against which to measure it. And what sense can we make of that possibility? How might we keep responsible a kind of hoping which by its nature is neither dependent on analogy from what we know to have been possible in the past, nor subject to the cumulative reality check of emerging events?

That, in relation to the climate crisis, seems to me to be the life-question of our time, and whether hope can mobilize the creative power to rescue a human future turns on a practically effective answer to it. The pivotal argument of this book is that we can indeed imagine a practice of responsibly realistic counter-empirical hope – but only when its permanent background is a recognition that the human condition is tragic. For that recognition serves for what might be called standing deference to the defeasibility of hope – which, framed by it, can then avoid facile forms of utopianism even while remaining resolutely counter-empirical. It does so not through ongoing contingent correction and qualification, which as we have seen it has to forgo, but by acknowledging as fundamental and permanent the feature of human life to which to Kant was pointing with his comment, noted previously, about no straight thing's ever being made out of the crooked timber of humanity. For a facile – that is, an irresponsible – utopianism is invariably one which sees no reason why its ideals cannot all be satisfied together, and aspires to straighten up the incorrigibly crooked human world accordingly. That is how it always comes to pursue more than the best that can be had. But the essence of recognizing tragedy as inescapable is accepting that irresolvably grievous value-conflict is inherent in human activity *as such*.

Towards a tragic vision

Tragedy as the term is intended here means not just any sufficiently drastic event involving death and mayhem – the loose journalistic misuse of the word in which it is routinely applied to fatalities arising from plane crashes, motorway pile-ups, gas explosions and the like. When the latest US massacre by a maniac with an over-the-counter assault rifle is described as tragic, however, we are closer to the proper sense. Behind these random individual events is the continuing plight of a society helplessly liable to such periodic obscenities because the 'frontier' values of individual self-reliance which are always being invoked in support of the Second Amendment lie at the core of that society's collective identity and elsewhere manifest themselves in various positive ways. That situation is tragic in the sense in which certain acknowledged works of drama and fiction are tragedies: it arises from something inherent in the key commitments of an agent – there an individual, here a society – such that the exercise of important strengths involves, by the same stroke of fate as it were, exposure to destructive forces.

To say that such situations characterize human activity as such is then to claim that in any real crunch, there are never reliably any win-win outcomes. The default human experience is win-lose, where gain in terms of some value involves loss, more or less grievous, in terms of one of the counter-values inevitably also in play. And such conflicting commitments are also characteristically *incommensurable*, so that there is no 'winning on balance', either way – just pain and gain to be enjoyed and endured together, and lived beyond (for the survivors) in unguessable ways. A tragic vision recognizes that we cannot 'put everything right', ever; where there are serious projects, we will always be exposed to failure, grief and pain. The core idea of the tragic is counter-utopian: standardly, not all the goods to which we are committed can be realized together. From this perspective, hope can always be invested in the potential for transformative change. But what it risks is not that experience will reveal it in this or that event to have been implausibly ambitious. Rather, its *standing* risk is that we are constantly drawn to embark ourselves in enterprises of aspiration which fail to pay due respect to our tragic nature. (In different terms, which the idea of tragedy in its literary forms can't help but suggest: the condition on which we exercise a capacity for creative action unique in nature is our permanent exposure to the danger of *hubris*, of trying to be more than human, a kind of failing – the attempt to escape the limitations of its species being – to which no other creature is liable.)

Why should we believe such deeply uncomfortable things about the human condition – especially now, after such an extended post-Enlightenment phase of trying very hard not to? In the first place, a tragic view can simply seem to offer the most persuasively honest framing for our experience as individual agents. The philosopher Christopher Hamilton begins a recent book with a forceful description of that experience which I can't do better than quote at some length:

> Human beings are born to suffer. All human lives are marked by pain and guilt, by loss and failure. … All of us go through life confused, and need in the end to acknowledge that life itself damages us, often profoundly and always irreparably. Human endeavour is fragile and human beings have only limited control of their own lives. … Our desires are often in conflict with each other, and our reason is a fragile instrument that is largely driven by blind urges and needs. … Most of the time we do not really understand what we are doing. … Human beings think they long for contentment, but when they get it they often destroy it because they are creatures who are deeply divided against themselves.

No thoughtful adult clear-sightedly reviewing his or her own history could, I take it, seriously quarrel with much of that. Nor can it readily be dismissed as the product of one of those dispirited moods which are (we all know) to be resisted by exercise, hard graft and the determined service of others. Hamilton in the same book very appositely quotes William James on this latter point: 'The method of averting one's attention from evil and living simply in the light of the good is splendid as long as it will work. ... But it breaks down impotently as soon as melancholy comes ... visions of horror are all drawn from the material of daily fact.' We certainly do not have to be held so captive by this view of life that we see nothing else, but unless we escape into denial we are surely bound to acknowledge it as our background condition, to be lived with as constructively as we can each manage.

Beyond its unflinching honesty, however, Hamilton's account also helps to make clear that what is at issue here is not just the everyday struggle which besets us because humans have evolved as complex creatures with many different wants and needs, and thus with multiple goals which often cut across one another. Mary Midgley notes in her indispensable *Beast and Man* how this evolutionary history gives us one sense in which conflict among normal desires and incompatibility between legitimate commitments is a deep-seated natural feature of the human form of life. Yet as she also remarks, we have an equally characteristic (and equally natural) drive towards self-integration, which often enables us to manage these clashes by compromise, sublimation and sacrifice. Adult human beings come ruefully to recognize – recognition being, indeed, a condition of adulthood – that we can hardly get by without daily battles for such balance. But compromise is not the stuff of tragedy – nor even is sacrifice, if compensated for by enough harmony among what remains. What Hamilton brings out by contrast is the core of the concept, something working at a level well below this ordinary diurnal churning: that is, our liability to find ourselves so deeply self-divided that human life – life realizing itself in human form – has to be seen as *irreparably self-damaging*.

It is this central core which plays such a large part in making the idea that our lives are inherently tragic so difficult to accept. Implicit in the very concept of life seems to be a positive nisus, an expectation that the normal condition for any life-form is a drive towards its characteristic kind of flourishing. Recognizing inevitable and irreparable self-damage as part of that flourishing involves making a decisive break with the problematizing utilitarianism which has been for so long the whole cast of thought or mental atmosphere of Western culture. It involves recognition that grief and suffering are not after all inimical to human flourishing, but instead are integral to it: they are an essential part of our living wholeness. That in turn requires us to see that the self-damage to which we are tragically exposed is a feature of our uniquely species-characteristic life in the realm of meaning and value.

Conflicted life-energy

It lies at the heart of our tragic exposure that, unlike all the other creatures who live unreflectingly in the flow of their differing natures, human beings are such that in them, different manifestations of life-energy are always coming into conflict with one another. That happens because we experience our own nature, through which our energies are expended, as giving rise to *imperatives* – requirements of which we are conscious *as* requirements, as deeply rooted commands governing our actions; and the inherent duality of our nature as rational animals means that these imperatives can be and often are encountered as commanding us contradictorily, in things which matter centrally to us. This is clearest in the case of moral requirement and the resistance which it is always meeting – resistance not so much from desire, which morality is precisely the command to overcome (recognizing that I *ought* to do something is recognizing that my merely not wanting to do it shouldn't count), as from claims which cannot be overridden but equally cannot be moralized.

A dynamic of universalization inherent in our rationality characterizes moral requirement. It takes its rise from our ability to be aware of motivations pulling us in different directions as presenting us with alternatives, and thereby with the primordial *why?* – why do this rather than that? As the philosopher John McDowell puts it, we should think of the natural course of an individual human being's intellectual and moral development as the acquisition of a *second nature*, essential to which is 'a kind of distancing of the agent from the practical tendencies that are part of what we might call his first nature'. This distancing process seems to enable one 'to step back from any motivational impulse one finds oneself subject to, and question its rational credentials'. Rationality and the distance which it gives us are inescapably distinctive aspects of our whole human nature. And the kind of credentials for which the practical *why?* is appealing when we step back will naturally be constituted by a *reason* for doing this rather than that, something which will have a second-order status more general than the particular alternative motivations – to address our difficulty, it will need to function as a rule of action: "In circumstances such as these, always do this rather than that."

But of course, this can't be the end of the matter for our rationality. Rules of action themselves come with attached motivations, because they are necessarily embedded in the practice of particular communities or constituencies of justification, within which we are each situated and to which we are more or less strongly inclined to conform. (It is, as we have learnt from Wittgenstein, in the very nature of a rule to be so embedded in a public practice, absent which there is ultimately no difference between privately *obeying* it and *deciding* individually what we shall take to be its demands as we go along – and thus, no rule at all.) Transcending this situatedness, however, is already built into the distancing reflex. For if my original motivation stands in need of rational credentialing, so also must any rule (whether of prudence or morality) which I reflectively adopt, since the weight that I am inclined to give different considerations in formulating it can equally be seen, in the 'distancing' perspective, as expressing something to which I 'find myself subject'. And carried forward by this rationalizing momentum, we will always be bringing more and more widely based rules of action appealed to in deliberation under the same kind of scrutiny. Once "Why do one thing rather than another?" moves via "Why do what accords with what I am reflectively inclined to adopt as a rule for this kind of case?" to "Why follow what is acknowledged as such a rule around here, by these and these people?", it has become clear that this momentum will push us into reflective distancing from *any* rule, embodied in the practice of any actual community of justification (why do what represents the prudential norm for our kind of society? The moral norm for the European heritage? … for Christian civilization? …) – until we reach a rule which is not open to such questioning.

Different ethical casts of mind have found such a stopping point in different places. For the Kantian, it is found in the rule "Always act rationally" – not, that is, in the sense of *rationally* where this means "Act so as most effectively to pursue whatever ends you happen to have", but rather: "Pursue acting rationally as its own self-evident and overriding end." This rule has what Kant thought of as *categorically imperative* force. Untrammelled by ifs or buts, it stops the questioning of rational credentials firmly in its tracks, because the question "*Why* act rationally?" is simply ill-formed – it assumes the practice of giving and acting on reasons which it therefore cannot call meaningfully into question: the rule "Act rationally" is thus the only rule which we are bound *in reason* to follow without being able to give any reason for doing so. Those of a consequentialist mindset, on the other hand, will find rational distancing coming to a natural end in some such rule as "Always promote happiness", in relation to which the question *why?* seems not so much ill-formed as quite inane – there being (self-evidently, as the consequentialist will think) no more general human motivation to be invoked in explanation. Either terminus, though, renders *universal* the relevant community for the rule comprising it. The only way to follow the rule "Act rationally", just as such – that is, a rule of action which invokes only the concept of being guided by reason – is to act in a way which no one could have any reason not to endorse, and that must mean acting in a way which does not treat as a special case or otherwise privilege anyone whom not everyone would agree so to privilege: in other words, acting with even-handed respect towards literally everyone's interests. And the rule to promote happiness, since its very ground of ultimacy is that happiness is everyone's fundamental desire, must in its unqualified form already express the thought that ultimately everyone's happiness is equally to be promoted. Thus it is really a perfectly consistent development out of our rational nature that prudential rules ("Do this rather than that in your own interests") find themselves leaning for support on moral rules ("… but compatibly with the interests of everyone around you") and moral rules find themselves claiming a universal constituency ("... and that means *literally* everyone").

The outcome of this development appears not just in the reach or extension of moral concern, but also in the seeming all-pervasiveness of moral reasoning – not only does no one seem to fall outwith the purview of our obligations of justice, honesty, benevolence and the rest, there is also no prompting to action which seems immune from this kind of consideration. That is the point of McDowell's claim that one can distance oneself in this distinctively evaluative way – in the mode of comparatively weighing one's motivations and bringing them under rules of action – from *any* motivational impulse to which one might be subject. But actually, that claim goes too far.

To see why, consider the motivation to behave *loyally* towards someone to whom I owe loyalty. In the 'distancing' which the 'stepping-back' reflex produces, I recognize my distinctness from my motivation, but always on the basis that I can step forward again and re-embark myself in this motivation, in order to be carried into action by it, should it be endorsed by deliberative 'weighing'. I can't, though, come back to reassume my loyalty after it has survived critical questioning, because *that* motivation can't survive it. Loyalty is of its nature *un*questioning, since a commitment to it for which I have to provide myself with reasons (never mind how compelling) will be as little to be trusted as the Vicar of Bray's successive submissions. To step back reflectively from loyalty is to have begun betraying the person to whom it was due. And loyalty is in this respect characteristic of a set of 'motivational impulses' which play an absolutely crucial role in the human form of life.

We could call these, for the sake of a general term to set against the imperative of morality, expressions of an imperative of *allegiance*. It is not that I cannot 'step back' from, for example, the motivation to favour my brother whenever favours are in question, just because he is my brother – I can certainly register this as an impulse to which I happen to be subject. But the process defeats itself right from the start as a rationally credentialing move. My motive here is not open to being acted on after reflective endorsement, because anything on which I so act won't be the motive of brotherliness. It is not brotherly to favour your brother at the behest of a rule, not even of the rule "Always favour your brother". Genuine brotherliness means favouring him in a spontaneous blood-response to him as your brother, and if you have got so far as to consult any code of conduct over this response, you have already failed in it. Similarly with: "On reflection, I find good reason to remain faithful to my wife" – the having seriously subjected conjugal commitment to that kind of scrutiny is already essentially faithless. Or again: if you look for a reason for loving your children, or your parents, or your ancestral place of settlement, *love* will not be what any reasons which you find will actually be rationalizing for you. The point here, one which McDowell's formulation quite fails to acknowledge, is that there are vital motivational impulses which reasoning about them doesn't tend to legitimize, but rather annihilates. These impulses, of which the natural force resides precisely in their *not* being placed at critical distance by the agent subject to them, are all essential in broadly the same way to our human being. Without the dispositions which are expressed spontaneously in kin-group loyalty, spousal solidarity, familial affection and *oikophilia* (Roger Scruton's useful coinage for love of your native place), to name but the most obvious, we could neither have emerged, nor could we long maintain ourselves, as this particular kind of social primate: language-using and creative of language-formed sociality, memory-reliant, practically collaborative, radically cultural and (in one important aspect of our nature) rational. They are, as one might say, the dispositions constitutive of the animal dimension of our rational animality.

It is also important to recognize that something very like allegiance can make a claim within the self. A determination to be true to what presents itself irresistibly as a life-purpose, a commitment around which the meaning of one's life finds itself organized, might well be characterized as a kind of loyalty – to one's destiny or one's star, as it were. This is just as constitutive of the human life-form as are the loyalties of other-directed allegiance, since it is the condition for a sense-making creature of its experience's falling into a pattern of significance around that given core. It is, further, a crucial lien of our sociality, since in its stronger manifestations it drives the leaders and creators whose individual vision is strong and clear enough to provide direction to the collaboratively sustained realm of meaning, rules, institutions and practices evoked in the previous chapter. And it is resistant just as much as are external loyalties to the rational reflex of 'stepping-back' – for the distancing move is, as we have seen, essentially an appeal to grounds on which some motivation could be *chosen* over others, whereas the crucial characteristic of what presents itself genuinely as a life-purpose is that it does so in a way which simply leaves one no choice.

By the same token, the dispositions of allegiance resist any universalization because they direct our concern *essentially* at particulars. Of faithfulness to my star this is plainly true – it is not followed as 'the kind of star which ...', because that would be offering grounds for choosing the essentially unchosen, and would involve not just self-betrayal but self-stultification. But nor are our other-directed loyalties moralizable into general applicability. If you try stepping back from the actual call of brotherhood, for instance, on the reflectively rationalizing track, what you reach is the kind of universal human solidarity which we try to have it both ways by calling *fraternity* – 'the brotherhood of man', with which, as the Australian philosopher Peter Singer eagerly puts it, 'the leaders of the French Revolution neatly conveyed the Enlightenment idea of extending to all mankind the concern that we ordinarily feel only for our kin'. But if all men are my brothers, no man is, since by definition a brother is someone to whom I stand in a distinctively close relation; and if no one is anyone's brother, the condition of universal fraternity is as empty as the French Revolution in fact demonstrated it to be. Meanwhile, of course, if I treat my actual brother as merely the representative nearest to hand of the claims of fraternity, not only is he likely to see and resent this, but he will be humanly quite warranted in doing so, since I shall have betrayed the real loyalty which I owe him. (That the idea of such blood-loyalty so embarrasses our rational-evaluative sensibilities only shows how pertinaciously the universalizing reflex of our second-nature seeks to dominate our practical thinking.)

Hence it is that human life is characteristically riven by strictly irresolvable conflict, at the behest of naturally arising imperatives which simply cannot be reconciled. They cannot, because neither a clash of allegiance with moral obligation, nor of one allegiance with another, nor of either obligation or loyalty with the call of one's star, can be rationally arbitrated – that is, compromised or negotiated by appeal to some more general principle – without disavowing and denaturing what is really at stake. Despite surface appearances, the *ought* which expresses the constitutive imperative of allegiance is not the *ought* of rational morality, where if I ought to do A, then generally *one ought*, if in my kind of situation, to do A; instead it represents something much closer to the etymological root – 'I *owe* him favour', from which nothing follows about what anyone else owes him or I owe anyone else. Assertions of allegiance put the rule-invoking *because* of reasoning aside, and simply redirect attention to the fact of embedded customary relations – in effect, they ask us to hear 'He's my brother' or 'She's my wife' in the right way, which would be to see how the question of special consideration to him or loyalty to her is thereby closed. All situations turning on this kind of issue are potentially tragic because in them the flow of life-energy through our nature is placed inescapably at odds with itself. Potential tends to become actual tragedy where characters are strong, demands are urgent and circumstances more malign than usual. These are the kind of conditions on which tragic drama and literature seize for exemplifying this profound feature of the human condition and symbolically addressing it. Thus a clash of allegiance with allegiance springs the classic dilemma of Agamemnon, whose loyalties to his followers as king and martial leader are brought cruelly into conflict with his kin loyalty to his daughter whom he must sacrifice to appease a capricious goddess. The conflict of allegiance with destiny is exemplified by Macbeth, whose feudal duty to Duncan struggles against his 'vaulting ambition', which represents the whole set of his ferocious nature as harped by supernatural visitation. Increasingly in the modern era, tragic conflict arises between allegiance and morality – the type-case here is Hamlet, who pledges his murdered father's ghost to 'sweep to my revenge' and then spends most of the play seeking moral certainty as to what the Ghost has told him ('I'll have grounds / More relative than this'). And more recently – keeping only to the great dramatists – we find the plays of Ibsen featuring a variety of clashes between loyalty to one's star and the expectations of rational morality. These themes are so powerfully present in literary tragedy, of course, just because the point of art is to help us understand and support the pervasiveness of tragic conflict within our actual lives.

Tragedy is thus endemic to human being. Life-energy expresses itself in living things as the species life-form realizing itself unimpededly in the particular individual life – a condition which is lived, and where consciousness has developed is experienced, as flourishing. Non-human conscious creatures we might think of as going happily with the grain of themselves when they are in this condition, but the dual-natured form of human life can only realize itself unimpededly, can only appear fully as itself, in the fundamentally conflicted ways which we have been examining – exposure to which therefore *is* flourishing in our case. Human beings flourish, wholly naturally, as rational-reflective universalizers *and* as constitutively embedded in patterns of sociality configured by allegiance *and* as unique individual sense-makers committed to the unchosen personal priorities without which they cannot make coherent sense. Necessarily therefore they flourish only through struggle, with its inevitabilities of grief and pain. This is what Nietzsche was pointing to when he identified the 'metaphysical solace which ... we derive from every true tragedy, the solace that in the ground of all things, and despite all changing appearances, life is indestructibly mighty and pleasurable ...'. For of course (though amid one's own individual toils this is always very hard to see), in any conflict of one manifestation of life-energy with another, what wins out in the end can only be life-energy. Here is a philosophical rationale for the insight which Blake puts with his matchless luminous simplicity:

> Man was made for joy and woe
> And when this we rightly know,
> Through the world we safely go.

But, crucially, *safely* here does not mean unharmed or undamaged – it means, without fear of losing our real identity. While 'damage' is by definition something which an ordinary object or living thing would be better off without, humans could not be better off without tragic conflictedness and its consequent suffering, because any being without those fundamental characteristics – any being which really thought of itself as always landing, or about to land, in Utopia – would not be human.

***

To summarize: hope against hope based in life-responsibility must still meet the condition of realism in order to count as hope at all. To prevent such hope from misleading us into magic thinking and a facile, irresponsible utopianism, to retain the defeasibility which is the necessary corollary of realism, transformative aspirations must recognize the normal human situation as tragic – a condition deeply rooted in our dual nature as rationally evaluative social primates. This recognition, reconnecting to a folk-wisdom with which we have latterly lost touch, means accepting the normal inevitability of grievous value-conflict in human affairs, such that the irresponsibly utopian aspiration to realize all our important commitments together becomes *inherently* defeasible.

And while we might seem, in these considerations, to have left the climate crisis some way behind, the whole point of course has been to return to it and foreground it once more in what will hopefully be a newly revealing light. To this task I now turn.

6

Climate Crisis as Tragedy

The account given in Chapter 5 of the underlying human condition as tragic arms us against the temptations of an irresponsible utopianism, but it also leads us towards a better grasp of the climate crisis. Certainly, if we are now urgently compelled to bring hope against hope to bear on that crisis, and such hope must be provided with a standing defence against those temptations if it is to do the proper work of hope instead of lapsing into wilful escapism, then a recognition of the tragic as central to human experience must be vital to that hope's defeasibility. Such recognition can also, however, give us important insights into the actual structure of our climate plight, which itself exhibits the characteristic tragic feature of life-energy coming into inevitable self-conflict.

A pattern of key strengths bringing with them exposure to destructive forces, which represented our initial scoping of the tragic, is indeed readily identifiable in the aetiology of climate emergency. The secular and instrumentally rational Enlightenment spirit which has produced so much worthwhile life-improvement across the world has also generated an apparent inability to rein in the relevant activities before they do irreversible harm. Distinctive human capacities which Western civilization in particular has realized – to make rational deliberated choices, to base belief on evidence and empirical testing, to free ourselves from ignorance, superstition and dogma – have been accompanied in their development by the striving for mastery and control which has betrayed us into doing decisive eco-systemic damage. Meanwhile the real and undeniable material improvements which exercising these capacities has brought us have blinded us (at first through ignorance, and latterly through various forms of denial) to the extent of the damage entailed. And now, to have any chance at all of escaping the worst consequences of that damage, we must at the very least severely constrain some of the key Enlightenment liberties currently being exercised in 'the pursuit of happiness'. Many of the classic ingredients of tragedy seem to be here. Moreover, the good and the evil in this situation are apparently incommensurable. *How much* present life-improvement, in terms not just of consumer durables but also of better health care, for instance, would be outweighed by *how much* future famine or drought? – any such question seems simply unanswerable, and not because the calculations involved would be very hard to do.

At this point, however, empirical realism is likely to be found retreating into the last ditch in its refusal to acknowledge hope against hope as our only hope. Surely, it will be said, these developments confront us with extraordinarily difficult *problems*, but to regard them as inherently intractable in the manner of a tragic conflict would be a betrayal of all that the last three centuries have taught us about our amazing resourcefulness and inventiveness, both technical and institutional. Even if it be admitted that the general human condition is tragically configured, that cannot mean that we are always helplessly conflicted in every aspect of our lives, which would seem to license a lazily resigned abandonment of every practical endeavour. We are not necessarily disabled by that tragic exposure from planning and successfully executing major projects – for example, engineering works, mass vaccination programmes and international rescue packages for various kinds of disaster. Can we not make sense of our climate plight as just the most challenging of the social and political difficulties which it is now our responsibility, somehow, to resolve on these lines?

We must give that response due consideration.

Problems, 'wicked' and 'super-wicked'

Ours is overwhelmingly a scientifically and technologically given age. If that remark sounds rather less trite than it used to, with the ascendancy of Orwellian 'alternative facts' in post-Brexit British politics and a bumptious scientific illiterate only recently ejected (for now, anyway) from the White House, it still points to a very fundamental feature of modern life and culture. Familiarly, that feature started to emerge during the seventeenth century, was entrenched by the Enlightenment and then drove the Industrial and subsequent technological revolutions to produce the overwhelmingly mechanized, urbanized and expertise-dependent world we know today.

Correspondingly, ours is an age of problems. That is so in the obvious sense that the mechanization and urbanization have brought with them, as well as many undoubted benefits, a whole raft of difficulties, dangers and forms of distress; climate change triggered by powering the successive technological revolutions with fossil carbon is perhaps only the most dramatic and all-embracing of these. But that we live in an age of problems is also true in the further sense that we characteristically see, and often seem only able to see, all those associated difficulties and distresses *as* problems.

This is perfectly thematic in relation to the techno-scientific mindset, with which the identification of any particular crux as a problem goes hand in hand. At least during periods of 'normal science', for any arena of understanding to claim scientific status its concepts must form a system which is generally well defined, so that apparent anomalies will offer themselves as requiring to be resolved in terms of those concepts. Etymologically, *pro-blema* derives from Greek *ballein*, to throw, so a problem is essentially something thrown forward, or as we might colloquially say, sticking out and requiring adjustment. Problems are wrinkles in some generally well-laid theoretical carpet and the obvious thing to do with them is to smooth them away, as far as we can. This yields both a strategy and an attitude of optimism: problems disturb the wider field, but we can hope to resolve them piecemeal just because they can be isolated vis-à-vis the system – as a matter of logic, *most* things in any given field can't be problematic, just as most things can't be abnormal. (And when any important problem vigorously resists such treatment, a paradigm shift may be in the offing which will inaugurate a new normal science with a different configuration of the unproblematic.)

Explicitly scientific thinking aside, however, the habit of identifying every form of difficulty or distress which we encounter as a problem, with the accompanying expectation that somehow, somewhere it will have a solution, is now very deeply ingrained in the contemporary approach to life. Individually, we arrive trailing clouds of problems with our feeding, teething and sleeping, which only intensify with our schooling; we then display all the usual problem behaviours in our teens, face the problems of earning a living and getting by as sexual beings in adulthood, and eventually decline through all the problems of old age until – sans everything – we find ourselves at grips with the one problem for which as yet no solution has seriously been proposed. (Though even here, the wishful proponents of cryogenics are at work.) En route, we take our pains as problems to the doctor, our marital difficulties similarly to the counsellor, and our sins (committed or contemplated) to the Problem Page. Collectively, we expect our politicians to solve our economic problems, our transport infrastructure problems, the problem of poverty, the problems of an overstretched NHS and so on across the whole domain of social policy. Or rather, since (given the kind of politician whom telly-democracy has largely produced) that expectation has probably now lapsed, we at least continue to berate them for failing to find solutions in any of these areas, and clamour for them to commission the necessary experts. The problem-solution template expresses the Spirit of the Age.

It is wholly unsurprising, therefore, that the dominant sustainability paradigm has represented a concerted attempt to mainstream environmental and climate issues by problematizing them, and thus framing them as accessible to social-scientific understanding and technocratic management. It has conceptualized human relations with the natural world as a series of in-principle-resolvable adverse human impacts on the general well functioning of the system comprising the human economy and the biosphere. The application of this approach to the issue of carbon emissions has then given us the overarching 'problem of climate change' – that is, for sustainability thinking, the problem of coping with the prospects and consequences of anthropogenic global warming so that it neither unduly disrupts the world's economic and social systems nor jeopardizes the defining contemporary human project of material progress. It has also given us an ever-expanding agenda of sub-problems – political, technical, institutional and social – including the problem of getting robust international agreement on emissions targets (leading into the sub-sub-problem of arbitrating differential equitable responsibilities as between the developed and developing nations); the problem of shifting on a large enough scale from fossil fuels to renewables in pursuit of these targets; the problem of finding, and making economically viable, techniques such as carbon capture and storage for taking out some of the CO_2 already in the atmosphere; the problem of devising appropriate incentives to change emissions-generating behaviours and patterns of living; the problem of educating people effectively about all these problems; and so on and on.

Thus readily does our thinking problematize its objects, up to and including an issue so complex and tangled as the climate crisis. But as that ramification of sub-problems and sub-sub-problems might have suggested, and as our continuing experience of trying to deal with them has surely demonstrated, the straightforward formulation of anthropogenic climate change and other environmental issues in terms of an agenda of problems has been very far from producing anything that could plausibly be called solutions. As a direct result, there has recently been increasing interest in trying to distinguish between ordinary problems and what have come to be called '*wicked* problems'. Problems can be wicked in the way a circle of argument can be vicious – their structural recalcitrance can easily affect the pursuer of solutions as having something malign about it. Among the main characteristics of wicked problems, according to the sociologist of science Steve Rayner who has developed the application of the idea to climate change in particular, are that:

- they are persistent and not amenable to clear definition, since different stakeholders with differing agendas are always involved;
- they offer little room for trial-and-error learning, because they typically implicate our life-arrangements so deeply that the situations presenting them cannot just be experimentally tweaked to see what might happen;
- they call forth 'contradictory certitudes' (that is, their core issues can be presented within alternative value-framings which conflict one with another); and thus,
- no clear-cut solutions are available (and attempts at solutions standardly give rise to new problems).

Once these features have been spelled out, it is apparent that not just climate change but a wide range of other environmental issues which we have become accustomed to approach within the broad sustainability paradigm, including water resource management, the use of GMOs in agriculture, waste disposal, marine ecosystem protection and biodiversity loss, answer readily to the 'wicked problem' description. And already in the climate change field, recent insightful criticism of the top-down global regime represented by the UNFCCC, the IPCC and the elaboration of machinery from the Kyoto Protocol through to the Paris accords has drawn on this conceptualization, pointing to the persistent failure of that approach adequately to recognize the tangled complexity and evaluative contestability of what faces us.

Indeed, it is on reflection apparent that the terms in which wicked problems are here characterized actually describe *most* policy issues. Complexity, resistant embeddedness and evaluative contestability evidently aren't absent from the problems of (for instance) poverty, hugely unequal wealth distribution, obesity or adolescent mental health in liberal-capitalist societies, nor from long-standing international disputes like the Israeli-Palestinian conflict, nor from the worldwide phenomenon of recrudescent religious fundamentalism. Once problematic 'wickedness' of this kind is explicitly acknowledged, that is, it can be seen to lurk everywhere, and the range of policy issues which are comparatively 'benign' narrows dramatically, until it includes essentially just 'engineering' matters (and then only until we recall HS2 or Heathrow expansion). The wicked problem, one is then tempted to say, is the *typical* policy challenge for our kind of diverse, techno-managerial and globally networked way of life.

Acknowledging this, recent commentators have sought to distinguish out the problem of climate change in particular by further specializing it as '*super*-wicked'. A super-wicked problem combines the characteristics of a merely wicked one with the additional features that:

- the time for solving it is running out (the problem will at some point so intensify as to become irreversible and unstoppable);
- those trying to solve it are also continuing to cause it (governments sign up to the Paris targets and plan airport expansion; most of us find ourselves driving to the recycling centre);
- no strong central authority is responsible for finding solutions (as the inadequacies of the UNFCCC process have illustrated all too clearly); and
- discounting anyway defers solutions constantly into the future (weighing definite present consumption against uncertain future benefits, people tend to choose what you would expect them to choose).

Again, once these features are identified we are likely to note that even as a super-wicked problem, climate change has at least one obvious parallel in the case of post-Cold War nuclear proliferation – which certainly exhibits the first and third of these attributes, and the world's experience of trying to deal with which is not a very happy portent. But possibly in some policy-academic forum the concept of the *extra*-super-wicked problem, designed to be instantiated exclusively by the climate crisis, is already under investigation.

If that speculation seems a bit flippant in the circumstances, any such tendency here surely registers a sense that the appearance of progressively refining a diagnostic intellectual tool isn't after all – however well intentioned those involved – fundamentally serious. Might the real point be, not increased theoretical clarity, but the more and more insistent occluding from view of any alternative to the problem-solution framing?

Problems and values

To explore that possibility further, we need to look briefly at the *value* dimension of wicked problems (in either of their degrees). All analysts are agreed that the defining 'wickedness', or malign recalcitrance, is supposed to be possessed by these problems in virtue of the plurality of value-framings and commitments which they call into play. A wicked problem cannot be thought of simply as a wrinkle in a smooth carpet, an interruption in the spread of a given value across a field of social concern, because in any such problem there are various different fields inter-tangled and a range of value-perspectives available. Climate change is identified as a wicked problem because, for instance, from a technocratic global-management perspective, rigorously policed emissions targets could help to combat rising global temperatures, but that would be at the cost of aspirations to a comfortable Western lifestyle for lots of people in India and China and elsewhere, something valued from the perspectives of both individual liberty and egalitarianism. Or again, substituting biofuels for fossil fuels would contribute to the global management goal of reducing CO_2 emissions, but only at the cost of taking up land which an egalitarian value-perspective would want to see used for food crops to alleviate global food poverty in the name of justice. The crucial point is that none of these perspectives is either obviously wrong or obviously right, nor is any of them analytically more fundamental than the others, but each offers simply a different socially configured spin on the data with distinctively different policy expectations to correspond: hence the 'wickedness' of any problem requiring their coordination or integration in order to solve it.

It follows thematically that as this profile of what Rayner calls the 'contradictory certitudes' characterizing wicked problems, has come to be appreciated, approaches to addressing these problems have come to be conceived of as 'clumsy solutions'. A response is *clumsy* in this sense to the extent that it does not adopt a single value-framing as its rationale, nor anticipate any clear-cut positive end-state, but rather appeals across the range of different evaluative constituencies involved. As a corollary, it will not rely on master-planning envisaging a well-defined goal, but on emergence and 'snowballing' effects to achieve results which constitute practical ways of dealing with the situation. In relation to climate change, an example would be the distributed, uncoordinated and speculative promotion of renewables, as advocated by Marco Verweij of the University of Bremen in contrast to the top-down (and evidently failing) global management strategies of Kyoto and after. Again, Prins and Rayner in their critique of the top-down global regime already cited, advocate the 'silver buckshot' approach (to distinguish it from the single, unachievable silver bullet). So instead of top-down targets premised on worldwide emissions trading and the Clean Development Mechanism, they look to an options-portfolio approach which combines building national and regional emissions markets from the bottom up, encouraging initiatives at state and city level, and in the business world opportunistically sector by sector, investing in R&D for both longer-term and stop-gap technologies and increased spending on adaptation, all with a much greater emphasis on a culture of learning and fine-tuning as we go. Mike Hulme describes the 'clumsy solutions' concept applied to this field as

> the idea of approaching the challenges of climate governance through a series of diverse, multi-level and almost deliberately overlapping, and even partly contradictory, institutions and policies. … By appealing to a wider variety of instincts and constituencies, such a bottom-up approach to climate governance may offer a greater prospect of delivery.

With a clumsy solution, that is, since everybody has the chance to get a bit of what they value along with some things they don't, the hope is that, unpredictably, convergence on a generally acceptable way forward (*feasible*, not *optimal* – because optimality assumes a single value-scale) may emerge.

Now all this can seem on the face of it persuasively loose-knit, flexible and postmodern. The trouble is, however, that the superficial value-pluralism is essentially a charade. The problem-solution framing of its very nature always implements an underlying value-monism.

We can see this most clearly if we try to prise apart the concepts of a *problem*, and of distress, discomfort or even just of things being other than one would wish. Clearly once *given* a problem, you have to seek a solution – that is by definition what you do with a problem; and similarly, if you are overridingly interested in solutions, problems must be what you expect to be confronted with. (With only a hammer, you are naturally on the lookout for nails.) By the same logic, when you suffer an *interruption* to your journey (for instance) you will want it to be as brief as possible, since again that is the attitude which suffering something as an interruption entails. Here it is apparent, however, that you don't have to regard every pause in every journey *as* an interruption: not all journeys are about getting as swiftly as possible from A to B, and there are halts (to relish the view, or to enjoy the sunshine) which for some journeys are the main point of the exercise. But suppose we try out the parallel thought that perhaps you don't have to regard every situation which is, even uncomfortably, not as you would wish it to be, as a *problem*. This thought will meet with very deep resistance, because finding ourselves in situations which are seriously far from how we should wish them to be seems necessarily to entail forms of *unhappiness*, and the idea that once you have unhappiness of any kind you have to minimize or ideally remove it, very readily seems to us basic and unchallengeable. "You don't have to regard every grief or distress as unhappiness" tends to strike us as merely perverse, if not as a contradiction in terms – which is why some form of utilitarianism, or duty of happiness-hunting, is the master-ethic of our age, and one with which the thought of any form of suffering as presenting ispo facto the problem of its mitigation or removal seems to go quite naturally.

It follows that a clumsy solution, for all its clumsiness, remains an approach to *resolving* a perceived clutch of difficulties. Or again, buckshot is certainly more effective than a single bullet if you don't know exactly where to aim (or maybe don't trust your aim), but the spirit of the metaphor is still that you are taking a pot at something hostile conceived of as standing across your path, in the hope if not of killing it then at least of sufficiently wounding or scaring it that it will lumber off. The single unifying parameter persistently in the background across all the undoubted value-plurality of 'wicked problems' is thus just the unquestioned Enlightenment value-framing of progressivism: every impediment to indefinitely making things materially better, or less bad, for human beings, even when it presents itself in the recalcitrant guise of a wicked problem, is an interruption to be overcome, a roadblock to be removed. The fundamental criterion of the onward and upward march of human betterment is thus always tacitly being appealed to, even by the option-rich, compromise-and-coping strategies which clumsy 'solutions' try to implement. Perception of problems as 'wicked' can helpfully diversify and sophisticate our practical approaches, but the master-value still, underlyingly, goes unchallenged.

But both wicked and super-wicked refinements of the standard problem-solution paradigm were supposed to be faithfully reflecting the strongly contested nature, the unignorable conflictedness presented by an issue like the climate crisis. If, at bottom, they are only masquerading as doing that, what are they really doing? What questions might they be trying to preclude us from asking, what inadmissible truths might they seek to shield us from seeing?

The alternative framing

I said earlier that ours is an age of problems, insofar as we seem only able to see difficulties and challenges, of almost any kind, on the problem-solution model which I have just been sketching. Putting it that way was certainly meant to suggest that such a predisposition could turn out to be a serious limitation. Were that to happen, of course, it would be despite its being that especially treacherous kind of limitation which inhibits *inter alia* our capacity to recognize it as such. (Compare the person who is so is so unassailably self-righteous that criticism, including of her self-righteousness, just washes off her.) But surely, we may by now be inclined to say, if anything is going to test the universal problem-paradigm to destruction and demonstrate where it is wanting, that must be the climate emergency. For it is quite hard *not* to see, even from the foregoing brief discussion, the strain under which our ideas of both *problem* and *solution* are being put in order to go on comprehending the climate crisis under this model. Correspondingly, it becomes increasingly hard not to recognize that reiterated upping of the analytical ante via wickedness and super-wickedness really represents, not refinement of our equipment for policy diagnosis, but a last-ditch defence of the model's favoured explanatory and practical status. The standard problem paradigm is surely being stretched just too implausibly far in this crucial case.

For what does the application of that paradigm to the climate emergency involve? The upshot of the foregoing is that three characteristic claims have to be made in order to sustain the belief that this emergency confronts us with any kind of *problem*. The first is that, however exacting the difficulties and dangers which we face, what we do by managing them in suitably improvisatory, option-rich and flexible fashion is to remain, albeit clumsily, in broad *control* of our situation. Secondly, this is achieved by compromising between and accommodating as far as possible the wide variety of contested values in play, while retaining them all intact as values from the differing perspectives involved – so that enough of the multiple stakeholders are kept sufficiently engaged on their own terms. And thirdly (the inescapable corollary of calling the upshot of these compromises any kind of *solution*), we thereby go on progressing, however conflictedly – the emergency will be contained, disaster avoided (perhaps very narrowly), the ring held, the trajectory of human betterment will go on slowly rising.

Suppose, though, that in face of the actual climate emergency as it presents itself to real, difficult experience, this is indeed too much of a stretch – suppose that none of these claims can really be believed to hold. That would mean recognizing the differing values involved to be *irresolvably* at odds, beyond any reconciliation however bodged up, so that there is no way out of the conflict between them without grief and pain in one direction or another. Moreover, any way through this conflict will involve suffering and disruption so serious that the opposed values themselves may not be able to remain unaffected – the whole value framework which has brought us to this pass may be shaken and destabilized. Hence we will necessarily lack any real control of our situation (since I only control a process which I steer in conformity to my unchanging values and desires: in any process during the course of which these are loosened or shifted unpredictably by the experience of the process itself, I am embarked but not in control). We will, indeed, be so far from being masters or managers of our climate and environmental plight that the very best we will be able to expect is somehow or other to avoid final catastrophe while saving and fostering what we can through a run of now-inevitable disasters. And with these truths, we are evidently on the terrain of tragedy as I began to outline it in the preceding chapter.

Tragedy is as we have already noted a concept from the humanities, referring canonically to literary forms, dramatic or narrative, which variously exhibit the profound inseparability of the grievous from the good in human affairs. Since these forms have been around for as long as there has been literature, we ought not to doubt that they express something persistent and fundamental in human being. They show us how our lives entail grievous loss and harm which events reveal to have been inevitable, given reasonable expectations about the chances of the world – inevitable because of the way in which exposure to radical conflict among our commitments is inherent in our situation. Whether this condition of always-potential conflict is represented, as often in classical Greek tragedy, as our being subject to different Olympian gods making different demands on us, or whether as in Shakespearian tragedy, a novel like *Anna Karenina* or the mature plays of Ibsen it is the bringing into play by circumstance of conflicting forces conceived as operating within the tragic hero or heroine, the upshot is the same: we find the material of tragedy in what Martha Nussbaum, in a famous discussion, called 'the everyday facts of lived practical reason'. Human beings are rational agents but also bring with them an inescapable heritage of allegiance; they are subject to competing and incommensurable imperatives and are often forced by circumstances into the dilemma of choosing between them; the upshot of these choices typically calls our precipitating commitments into question, and thus we always, at bottom, lack effective control over how our lives turn out. Using the idea of tragedy in this way to understand ourselves clearly runs at a tangent to any social-scientific thinking about the human condition. But this is the point at which to insist that the humanities in general have at least as much claim as has any method modelled on empirical science to explicate human existence, in its sociopolitical aspects as in any others. This is likely to appear initially to the modern mind as an alien thought, although it would not have done so at any time up to the late nineteenth century, when our grasp of our social nature started to become (scientistically) scientized. We should, however, try such explication out again, and very urgently – because if approaching the human condition social-scientifically *occludes* the tragic in the problematizing way which we have seen, and yet there are indeed tragic dilemmas in real political and social life, then *not to be able to see them* is going to make our plight as they come upon us simply impossible to deal with.

But it is clear that, brought to bear on the climate and ecological emergency, anything like a tragic view of the human condition has very uncomfortable implications. For that condition is now inextricably bound up with the fate of life at large, and the integrity of the biosphere and the habitability of large tracts of the planet are coming rapidly into jeopardy as a consequence. Accepting that in struggling through these conflicts we lack any real control over outcomes which are bound to be in one way or another grievous, would seem to imply in this context that humanity's sooner or later wrecking the Earth might have a tragic inevitability about it. Why should we nevertheless embrace such an understanding of our climate plight?

Climate and conflicted life-energy

There are two levels at which that question may be answered. The first involves showing that the patterns of human action and interaction which have generated and are now trying to deal with the climate emergency are tragically configured – they bring different human manifestations of life-energy into irresolvable conflict. But at the second, deeper level, we can also see climate crisis as life-energy itself falling into self-conflict. At either level, only a tragic understanding is adequate to the situation in which we are caught up.

At the first of these levels, our climate and ecological plight presents us with the most demandingly rationalized obligations under which human beings have ever been placed – to protect living things who (and which) don't yet exist, in a future which can only be represented through scientific modelling – and those obligations find themselves in conflict with very deeply embedded allegiances. Tragedy sits at the very heart of the climate emergency, in the conflict between on the one hand the claims of future generations, and also of non-human species, and on the other hand, demands arising from the way we are all more or less embedded in the contemporary economic and social forms now putting those beings in such jeopardy. If this could be seen as a clash between judicious fair-dealing and mere short-sighted selfishness, or between an ecocentric and an anthropocentric attitude to the Earth's living systems, or even between responsible plain living and attachment to a meretricious consumerism, it is easy to see why it would get formulated as a wicked or a super-wicked problem: pervasive, frighteningly hard to get to grips with, presenting a hugely daunting challenge to policymakers in every kind of polity from liberal democracies to command economies – yet still, not in principle irresolvable. But once we recognize the pattern of allegiances conflicting with moral imperatives as outcropping again here, the lights change. For the calls of justice are those of our natural inclinations to fair-dealing with our compeers, rationalized and universalized, while people are tied so very firmly into destructive and climate-jeopardizing lifestyles and habits very largely by the loyalties and demands of allegiance – loyalties to their communities and work-colleagues as they find them, and to their loved ones maybe living at a distance; demands for creative self-expression in whatever mode of activity their life-chances have made available to them, and for active human connection however many carbon emissions go with it. That because of the reach of ecological and atmospheric circumstance these attachments are in the larger view evidently destructive doesn't prevent them from claiming and holding us in a way that is deeply prior to the rationally evaluable. Thus the opposition between environmental responsibility as an insistent demand and material welfare as an economic and social driving force is made thoroughly intractable.

Again, as one increasingly urgent consequence of environmental crisis there is developing an ugly clash between on the one hand the responsibility of rich countries in temperate climes like those of Northern Europe towards rapidly growing numbers of environmental refugees, and more generally for the relief of increasingly frequent climate-driven disasters in the global South, and on the other hand the evident strong desire of established citizens of these temperate countries to look first to their own and preserve their cultural and ethnic landscapes to a large extent as they have known them. It is even easier to simplify this conflict by ethicizing it into a problem – in solving which there can seem little enough to be said for the values of nationalism or cultural conservatism as against those of charity, generosity and the saving of lives. But here too beneath the surface run centrally constitutive allegiances – those to one's own kin and community, and that of *oikophilia* or attachment to one's actual familiar place – which give to apparently dismissible concerns the occluded force of a quite undismissible demand meeting any ethical imperative head-on.

The imperative of life-purpose, as well as variously of allegiance, is also closely implicated in climate crisis. A commitment to progressively improving people's material conditions (in the broadest sense of the comfort, amenity and security of their actual daily lives) is an expression of important virtues – care, compassion, justice predominant among them – seconded by the technological rationality, practical imagination and foresight which human beings have developed with an ever-growing momentum since the Industrial Revolution. Such progress has vastly improved the human condition in a thousand ways, and the impetus to do so was admirable and for a long while perfectly reasonable. This pursuit of material progress has, however, acquired in modern societies its own perverted dynamic. The process of meeting ever-growing material aspirations has come to seem as if it could supply the purpose and meaning which we each need to make sense of our own individual lives. Hence the modern consumer's literal addiction to commodities, the unceasing competitive churning of which according to the inner logic of capitalism seems to offer an *ersatz* directedness to lives characteristically lacking it. This progressivism, as we may call it, is now fully developed as the religion-substitute classically formulated in the mid-nineteenth century by John Stuart Mill:

> Let it be remembered that if the individual life is short, the life of the human species is not short; its indefinite duration is practically equivalent to endlessness; and being combined with infinite capability of improvement, it offers to the imagination and sympathies a large enough object to satisfy … grandeur of aspiration.

Since Mill wrote, however, our vastly extended technological reach has caused the drive for present material betterment to assume forms which drastically imperil the prospects for indefinite future betterment or indeed survival. And this is a tragic dilemma because the future reference of material progress is essential – betterment *is* only progress if it doesn't at the same time undermine its own continuance. Of course, the sustainability paradigm was supposed to let us off this hook, but all it ever really offered was a way of sacrificing the future to the present more discreetly. Our combination of great and *actual* technical power with circumscribed, patchy and conditional foresight as to the effects of its use was always going to make the present use of that power for material benefit effectively incumbent on us – unless we could demonstrate with a robustness which the numbers just don't have that the cost of any particular present benefit was going to be intolerable, not just to the future but also to us contemplating it now as our legacy to the future. (On the precautionary principle, indeed, just the outside chance of a catastrophic upshot should be enough to constrain us without the need for such demonstration, which is precisely why the dynamic of progressivism has always ruled out the general adoption of this principle.) This is surely how we should understand the claim that 'sustainable development' was always going to fail as a template for policy: given the overwhelming investment of present people in progressivism as substitute life-purpose, it was always tragically inevitable that it would fail.

Beneath these tangled commitments across the human scene, however, we should also be prepared to hear the tragic ground-bass of climate crisis as a deep conflict of life *within itself*.

Life is always pouring out its energy into particularity – into the almost unimaginable differentiation of habitats and species, their vigorously proliferating forms of ecosystemic interdependence, the widely varying characteristics of races and lineages, the multiple bonding patterns of kinship and locality attachments, all the way to the distinctive gifts, powers and dispositions of each living individual. And with these particularities it also informs, quite naturally and thematically as one life-manifestation among others, the species-specific capacities of human reason. But it thereby launches against itself the universalizing drive which we have seen to inhere in these capacities and nowhere else in nature. For the universal comes only with conscious reflection, with the comparative bringing of particulars under *concepts* and of motivations under *rules*, and only humans do these things. And as humans have come to be the overwhelmingly dominant force on Earth – their tendencies no longer held in check by deficiency of power but vastly augmented by the release of fossil energy – that drive seeks more and more to rationalize and subsume the forms of particularity with which it finds itself confronted. It has been a familiar trope in environmental writing since the 1970s – see, for instance, Barry Commoner's still valuable book *The Closing Circle* – that human rationality, bringing with it conscious goal-directedness and the associated logic of linearity, has intervened as an increasingly destructive force in the naturally cyclic, feedback-regulated processes of the biosphere. But more is involved here than just overfishing the oceans and depleting the soil, logging the rainforest for roads or for palm-oil plantations, and generally concreting over the wilderness. There is also the still deadlier overlaying of human variety and diversity with the universalizing rational morality which yields progressivism: the conviction that since all human beings share essentially the same claims and needs, the only acceptable social goal at bottom is that everything should get measurably (and hence, materially) better for everyone, everywhere, indefinitely. And when that becomes, as we have seen, a substitute for life-purpose, all the other resource-exploitative linearities are massively reinforced. That has been the inner history of environmental damage since the Industrial Revolution.

While an awareness of something like this dynamic has been around for a good while, however, the dominance hitherto of the 'problem paradigm' has precluded any clear recognition of its real nature. But we should now be in a position to appreciate the full reach of this recognition. It is not just that the human condition is inherently tragic: further, the situation of life on Earth is itself tragic, insofar as it has led up through its natural unfolding to the introduction of the human species. Being themselves one particular manifestation of life's abounding joy in particularity, humans bring with them both the capacity and what has latterly become the urgent tendency to override and degrade all this vibrant variety, and now on a planet-wide, biosphere-jeopardizing scale. The condition of Gaia herself, we might borrow James Lovelock's rich metaphor to suggest, is tragic. Or if that metaphor is rejected, one might say instead that the Anthropocene marks not just a putative geological epoch, period or era, but the playing-out of a deep-lying tragic pattern in the natural-historical configuration of the whole Earth system.

For some readers this last may very well be a thought too far. But it indicates what I would contend to be the order of consideration with which we must be prepared to engage, if we want to see the climate crisis in all its true dimensions. In comparison, talking about this crisis in terms of problems and solutions is a cheap evasion. That does not call on us to yield up activism in despair – the association of a tragic vision with that kind of fatalism is entirely misleading, as I have tried to make clear. But it must, if my argument in these two chapters has been anything like right, very profoundly affect the quality of our hope.

Climate tragedy and rough coping

If a tragic understanding encourages us to see the climate crisis in these ways, that must clearly have significant consequences for the approach we bring to it – and it will be in the recognition of these consequences that the standing defeasibility of our counter-empirical hoping must consist. Rejection of the hunt for solutions as a general template for action must go with a decisive shifting in other related attitudes and approaches. Nietzsche remarked that the most important feature of what he called a tragic culture 'lies in putting wisdom in place of science as the highest goal'. That of course does not mean for us abandoning science – how could we abandon something without which we could not even *identify* the climate crisis? – but it points to an overall attitude reconfigured in crucial respects from our currently dominant cast of mind. Within this reconfiguration, scientific understanding can then take its rightful subordinate place, regarded neither as the primary mode of access to reality nor as our necessary fall-back provider of metrics for practical success. The sort of bind in which climate crisis catches us calls neither for solving nor resolving (however clumsily), but much more for something which might be labelled *rough coping*. 'Rough' here carries the sense not just of improvised approximation (as when we rough out a plan for getting through), but also that of expecting a bumpy ride with many breakages en route, and of being prepared to take that rough with whatever smooth may be available. It also acknowledges this as the general nature of our dealings with unvarnished reality ('life in the rough'). We are engaged in rough coping whenever we accept that we haven't failed because we couldn't have succeeded, that the best-laid plans can change daily (this is sometimes called resilience) and that with luck, life can offer us a choice between evils – though we shall still be left to *make* that choice, in climate crisis as in so much else.

Related to this last point is the standing acceptance, also, that pursuit of any relevant value – justice, say, between or within generations – will come at the cost of failing to live up to some equally relevant and powerful claim, such as those of full political liberty as many of us have been used to it, or compassion for the 'underprivileged' (or indeed the overprivileged), or tolerance of the technically harmless in lifestyles which will nevertheless have to be uncomfortably changed. This is the fundamental anti-utopianism, or at least the constraint on irresponsible utopianism, which comes with tragic vision. It involves not just becoming fully aware that we cannot ever 'put everything right', but actually cultivating a hard readiness for the necessary choices between different orders of wrong if we are to go on (roughly) coping.

It will involve recognizing also that the grievousness of such choices may change our accepted values in ways which we can't, from this side of the experience, really predict. What, for instance, might justice actually *come to mean* in a broken world – a world full of what Tim Mulgan calls 'survival bottlenecks', situations in which there are more people than there are resources available to support them in meeting their basic needs? In that kind of world, some value like justice would be needed to organize whatever institutional and regulatory mechanisms were to determine not simply 'who gets what' (the classic distribution issue in political theory) but *who lives and dies*, and the only such mechanism seemingly acceptable to the universalist conception of justice to which we have been accustomed would be some form of lottery – where though there must be losers, everyone has an equal chance of surviving. Plainly, however, no one is in practice going to accept that. ("If justice requires me to put my kids into a survival lottery along with a lot of strangers, then to hell with justice ... .") So how *will* whatever we can then regard as fair dealing (a concern for which, any kind of civilization will always need) sit alongside considerations of family and community loyalty? We are evidently going to have to trust, here and elsewhere in many similar difficulties, to a tragically-pressured inventiveness of moral creativity as part of the painfulness of rough coping.

A further important element of tragic vision here is acceptance that the climate crisis is nobody's fault. No one is to blame for secularization, aspirations to progress as a meaning-substitute or attempts to fudge limits with 'sustainability' constraints, any more than anyone is to blame in any genuinely tragic drama. (Who or what is to blame for the accumulating horrors in *Macbeth*? – Macbeth himself? his fiend-like queen? the witches? the compliance of Banquo? the martial culture of eleventh-century Scotland?) The notion of blame, offspring of morality and causality, is just an ungainly intruder at the level at which destructiveness is locked into these gathering actions. Certainly, in the climate case as much as in the play, people (and institutions) caught up in the process do engage in blameworthy behaviour – for instance, corporate malpractice in the suppression of evidence by the fossil-fuel industry, or breach of trust by politicians operating within the sustainability paradigm. But the tragic situation as a whole cannot be laid at anyone's door – so to talk as some do of the losses of species and ecosystems through climate change as something equivalent to murder is hysterical, albeit understandable. As the arguments of Chapter 4 should have prepared us to recognize, global biocide is a crime which we can't *pin on* anyone except the carbon-profligate majority of humanity – frustrating as this is for those looking to identify the punishable culpability which goes so readily with a moralizing discourse. But that is far from meaning that very sharp measures are not to be taken against institutions and individuals involved at key junctures in its occurrence. The question is not who is to blame, but who is in the way. Not being to blame for the emergency does not remove careless corporations, greedy executives, lying demagogues, trivializing journalists or any of a whole crew of variously adverse others from our sights. Since blame is beside the point, in fact, we need not worry about difficulties in attributing it fairly, but can simply aim to push aside (which of course does not imply eliminating, only rendering harmless) anyone or anything impeding drastic transformative action – a necessary ruthlessness which is itself one more painful dimension of the overall tragedy.

That ruthlessness, it must be said, also has to go for our dealings with ourselves, as agents of active hope in this bleak context. At the core of the approach which I have just sketched must be an unsentimental and unsparing focus on essentials, rather of the kind captured in this bitterly plausible prediction by Derrick Jensen:

> The humans that come after are not going to give a shit about whether we were pacifists or not. … They're not going to care if we recycled. They're not going to care about any of that stuff. What they're going to care about is whether they can breathe the air and drink the water. We're fighting for the life of the planet here.

***

To situate this chapter, in conclusion, within the argument of the whole book so far: hoping realistically in face of the climate emergency cannot be a matter of deriving precedents from past emergencies for what we can now achieve, since honest attention to any such precedents can only dismay and daunt us. But nor can it just be a matter of recognizing that human creativity can always generate unprecedented transformative possibilities, since that recognition alone could betray us, under the inevitable pressures, into utopian fantasy. Genuine realism depends on recognizing, too, that our creativity always operates within, and must be configured and constrained by, the necessary structure of human agency which a tragic view of life expresses. And the climate emergency, once it too is seen in this light rather than routinely and implausibly problematized as it has been for too long, shows itself as apt to be tackled in a spirit of tragically informed hope.

Strong in those recognitions, those who are working for a human future can maintain themselves unexhilarated but undefeated, as they turn their minds to the hugely demanding question of ways and means, with which the next three chapters engage.

7

On the Way to Revolution

Faithfulness to the life within us insists on our going on hoping. And responsible hope, conscious of anything much more than the immediate daily horizon of getting by, must now reach out towards the transformative changes in economy, society, politics and life-arrangements generally, which might even yet retrieve some kind of habitable world from the unprecedented perils now confronting humanity. But the burden of this book so far has been to argue that, just because the perils *are* so utterly unprecedented, transformative hope has to arise from depths of counter-empirical resolve at which it comes conceptually bound up with a tragic view of life as the condition of its necessary realism. I will not defend this claim any further, but will take it as having been made sufficiently compelling for the purposes of argument – although of course, what we think about the implications of accepting it will itself feed back into final assessment of the case as a whole.

We must now consider what those implications actually are, for the clamorous practicalities which beset us on every side. What can we take forward, into political and social action over the next two make-or-break decades, from the vital connections which I have been seeking to establish between hope, realism and tragedy? What difference would framing life-hope explicitly within a tragic vision make to our understanding of how we must grapple with our dangerous prospects, and seize upon our remaining opportunities? How would being properly realistic about the climate emergency *feel*, and where might it take us?

Here is the place to emphasize again a point made in the Introduction, that the answers to these questions and the consequent practical recourses which I suggest in the remaining chapters are all intended to be provisional and heuristic. The argument thus far, from lost empirical hope to life-hope informed by a tragic vision, is I think watertight – or at any rate, it is one which I myself find wholly convincing. By comparison, what follows explores forward, open-endedly, along lines which the conclusion of that argument seems to me to indicate. The proposals turned up by that exploration will, I am sure, be found starkly controversial by many. But if I have tried to state them as forthrightly as I can, that is not in any spirit of certitude, still less of rationalizing commitments already dogmatically embraced: it is an attempt at the kind of clarity which will challenge debate. *Something* along these lines seems to me inevitable – and we are now imperilling the human future unless we are discussing, honestly and vigorously, with no holds barred and no assumptions ruled out or in, just precisely what.

To start with, one very obvious difference which a tragic vision ought to make is to discourage us from easy simplification. A recent book claims for instance that 'Climate change … is domination. … It is the most recent manifestation of how powerful men throughout history have sought to steal from the less powerful and dismiss them as merely inconvenient. … Understanding climate change in this way transforms everything'. This is not purely idiosyncratic – as well as the occasional academic, one finds even otherwise sensible activists who appear to believe similar things. It is, however, a palpable misrepresentation; it reduces to the tokens of political correctness a multifarious reality in which, as we have seen, good and ill are inextricably interwoven and the apportioning of blame makes very little sense. Transforming everything, or anything, by abdicating from hard thinking will certainly not serve us. But nor, regrettably, is that sort of offering the comparative outlier that one might have expected. Here is a very well-known green-political figure, Jonathon Porritt, on why governments failed to grip the climate change issue in the late 1980s and 1990s: 'What really stood in our way, and still does today, is a powerful, ruthless, self-serving political elite that will brook no barriers to further enriching itself at the expense of the whole of the rest of humankind.' It shouldn't actually require that much tragic realism to recognize in neo-liberalism not some elite conspiracy, but a much more widespread attitude towards life. A tragic vision, however, would at least serve as a strong prophylactic against the temptation thus to erase complexity.

There are, though, more serious contenders by way of action-directed thought about our plight, and I want in this chapter to explore how three of them look in the light of such a tragically informed realism. This should help both to clarify what such realism involves, and to indicate how much further than any of these approaches we shall need to go.

Tragic resilience

What exactly does a tragic vision, or general habit of thought, itself consist in, when applied to practical issues? As a first stab at answering this question, it may be helpful to consider how close to the tragic the attitude and approach of 'Deep Adaptation' comes. This concept has recently been put into circulation by the sustainability management academic Jem Bendell, in an online paper which has apparently set records for the number of times it has been downloaded. The paper is a serious attempt, clearly striking a chord with many concerned people, to take a much bleaker view of what is coming by way of climate emergency than any official line consonant with the sustainable development paradigm or the Paris Agreement, and to note some of what would follow in practice from so doing. Does it offer an example of the necessary kind of realism?

Ongoing adaptation needs to be 'deep', in Bendell's view, to distinguish it from what the notion of adaptation to climate change has hitherto been taken to mean in management and policy studies. Broadly, that has been: strategies to keep our current (Westernized) societies functioning in essentially their current forms through climate perturbations held to be real and inevitable, but also in principle manageable. Adaptation so understood means political and institutional flexibility to ensure that sustainability – conceived of as an ecologized version of continuing progress – can bend before the strengthening winds of change without shattering or suffering any too drastic damage. Bendell doesn't go quite so far as to disparage this sustainability model as such: he doesn't call it, as one well might, the model on which we still basically have it all while pretending that this is compatible with future generations' having it all too. But he does see that what we must now adapt to is the ending of this and similar aspirations, rather than the reduction or even the quite stringent narrowing of our options for carrying on with them. That in turn means reading the much-favoured term *resilience* differently, and the relevant difference suggests why the adjective *deep* might at first seem apposite for this new kind of adaptation. Rather than thinking of resilience in the cheerily upbeat manner of the Stockholm Resilience Centre – 'the capacity of a system, be it an individual, a forest, a city or an economy, to deal with change and continue to develop' – Bendell suggests a less systems-theoretical and more psychologically searching conception: resilience is 'the process of adapting well in the face of adversity, trauma, tragedy, threats or significant sources of stress. … It means "bouncing back" from difficult experiences.'

But while this would-be deeper understanding references tragedy among our other challenges – and indeed, the whole paper has 'Tragedy' in its subtitle – the language in which tragedy is invoked decisively undermines itself, and thus at the same time the whole conception, and it does so in a very characteristically modern way. (The definition quoted comes, as one might have guessed, from a publication by the American Psychological Association.) No acceptable response to tragedy could possibly be formulated in those terms by anyone having the remotest real acquaintance with the tragic as a cultural phenomenon. It is not just that you could only 'bounce back' from (say) watching a performance of *King Lear* if you had not been properly attending; any 'creative reinterpretation of identity and priorities' (which is Bendell's gloss on such rebounding), in response to that or any other genuinely tragic work or experience, could have nothing at all in common with complaisant accommodation or the recovery of psychological equilibrium. The whole point about adversity, trauma or 'stress' affecting us at the level of tragedy is that it recasts our lives in ways from the grievousness of which there is neither recovery nor imaginable return – but of which we cannot sum up the outcomes, positively or negatively, either beforehand or in retrospect, because central to what is recast is our structure of values itself.

The resilience which we must cultivate as part of bringing a properly tragic vision to bear on the climate emergency, therefore, can be indicated by trying to supply what Bendell's psychological account omits – which means, trying to take the notion down to an even deeper, existential level. This kind of resilience means something much more like the ability to recognize that our well-being, security and even our identity (individual or social) are not in our hands because irresolvable conflict between commitments is endemic to our condition, but that grievousness is not the sole end of that story. It is the robustness required to go forward knowing that whatever we attempt, we shall fail in crucial respects, but also that we cannot know in advance *which* respects nor, correspondingly, where we will succeed (or at least fail less badly); nor can we know how that failure (along with any partial success) will change the way in which we look back at what we have come through, and feel about where we have arrived. For this is why there are necessarily no win-wins within a tragic understanding: serious tragic experience may be expected to shake up our whole structure of value so vigorously that the comparative evaluation of outcomes which might talk in terms of winning or losing – never mind of 'bouncing back' – cannot be preserved across it.

This is not just a deeper, but in our culture a now deeply unfamiliar cast of mind. We shall perhaps only grasp it fully when we have worked through what it means for tackling the issues which confront us. Meanwhile, we can note how the other components of Bendell's 'Deep Adaptation', which he identifies as *relinquishment* and *restoration*, need to be correspondingly re-thought. *Relinquishment* he describes as 'people and communities letting go of certain assets, behaviours and beliefs where retaining them could make matters worse'. This is obviously going to be an increasingly urgent requirement, whether we are thinking in terms of withdrawing from indefensible coastlines or abandoning unsustainable consumerist expectations. But where it will really bite within the frame of a tragic understanding is in relation to currently accepted norms and values (which, for Bendell, 'Deep Adaptation' is ultimately about maintaining). Tragic relinquishment means being prepared to hold values like justice or equality heuristically, in the way illustrated in the previous chapter – to go forward in the expectation that carrying them through to the other side of climate chaos (and the string of now-inevitable disasters which will attend even the best-case scenario) will change, perhaps out of present recognition, both them and ourselves. By the same token, *restoration* means for Bendell the rediscovery of attitudes and approaches to life with which high-carbon civilization has lost touch. This in his paper reintroduces the long-standing Green agenda of rewilding, dietary change and socio-economic localization – again, all of which will indeed be needed as the climate emergency presses upon us. But the key 'attitude' which we shall have to recover, and one which the abiding Green tendency to utopianism has actually helped to suppress, is just the vision of life configured in terms of genuinely tragic resilience, relinquishment and upshot as above.

In summary, the contrast with 'Deep Adaptation' highlights two components of a tragic mind-set which emerge from my argument up to this point, and distinguish that attitude clearly from other ways of reining back on facile optimism. In the first place, it recognizes the permanent liability of human experience to generate irresolvable conflict among commitments, and the human condition therefore as inherently rather than just contingently grievous. (As Hamilton has it, 'life itself damages us … always irreparably'.) It also recognizes, however, that the values and commitments, attachment to which yields tragic grief, are not static but subject to recasting and reinvention under tragic pressure, so that such pressure can itself have a creative dimension: the irreparable is not necessarily the merely destructive. The question is then how a tragic understanding in those terms might help us to negotiate our present plight.

Transformation as revolution?

One thing to which it might assist us is full acknowledgement that, as I suggested in the Introduction, 'transformative change' – a notion much invoked in current discussions of the climate situation, and general enough just in those terms to sound a broadly positive note of (perhaps slightly apprehensive) excitement for many – must actually mean social and political *revolution*: a term with a harsher and more fearful timbre. Social and economic transformation in itself does not seem to be incompatible with utopianism – all utopias, after all, have been presented as transformative developments from a non-ideal present context. But a utopian revolution is surely unimaginable; anything that could count as a revolution would have to throw too many comfortable things up in the air for that.

'Revolution', however, is a contested concept even within political theory, and beyond that domain the endemic restlessness of late capitalism has accustomed us to a wide variety of looser usages: so we have revolutions in fashion tailoring and in digital technology every few years, and in aspects of sexual and other forms of interpersonal behaviour at only slightly less frequent intervals. Equally, any significant shift in sociopolitical arrangements is likely to find itself styled revolutionary, however remote its ethos from Committees of Public Safety or Red Terror. So the question with which a tragic view might help us get to grips is: what kind of revolution do we now have to be talking about?

This question presses when we register that use of the term 'revolution' here need not be driven by the more or less passionate kinds of ideological commitment which typically motivated earlier revolutionaries. Rather, it can represent the careful consensus of relevant scientific expertise. Naomi Klein, in a recent book, calls on the testimony of the leading climate scientist Kevin Anderson from whom we have already heard in Chapter 1:

> Perhaps … at the turn of the millennium, 2°C levels of mitigation could have been achieved through significant *evolutionary changes within the political and economic hegemony*. But climate change is a cumulative issue! … Our ongoing and collective carbon profligacy has squandered any opportunity for the 'evolutionary change' afforded by our earlier (and larger) 2°C carbon budget. Today, after two decades of bluff and lies, the remaining 2°C budget demands *revolutionary change to the political and economic hegemony*.

(The emphases here are Anderson's own.) In other words, the merely objective facts about the size of our remaining carbon budget – the total volume to which we must constrain emissions for a chance of keeping global overheating below levels liable to trigger runaway intensification towards catastrophe – indicate unmistakably the need for very swift and dramatic interventions indeed. Specific policies may be argued over, but the *scale* and *reach* of such interventions can be suggested by the combination of: a complete moratorium on fossil-fuel extraction; effective carbon rationing to eliminate fossil-fuel use in favour of renewables over a decade at most; a consequent decisive rebalancing of the economy with radically changed patterns of employment; land conscription for local food security; and perhaps also the introduction of a citizen's income to ensure that economic shake-out need mean no one actually starving. These shifts, and others to similar effect, are so drastic and would need to be carried through in such direct defiance of vested interests currently so powerful and predominant, that only a revolutionary impetus, one would suppose, could give them any chance of coming to pass. But, again, what might now count as revolutionary in that sense?

If the answer to that question is after all just 'far-reaching social and economic changes achieved against stiff opposition from corporations and their political cheerleaders', then it could be claimed that we have two working models of revolution already on the stocks. These are the Green New Deal which has begun to gain traction on both sides of the Atlantic; and the Extinction Rebellion movement which took off so swiftly worldwide after it first surfaced in Britain in 2019. A look at each of these in turn should help to sharpen the issues around the ideas of revolution and tragic realism in this context.

The Green New Deal

This is essentially a rainbow alliance of climate campaigners, left-of-centre political parties, unions and 'frontline communities', assembled within a climate-justice framework and aiming to gain power through the existing electoral systems of the leading capitalist states. That power would then be deployed to reintroduce state regulation of capital and finance and direction of national (leading rapidly to international) efforts in the direction of net carbon neutrality. Target dates for this vary slightly between different models, but are much more ambitious than the mid-century touted by, for example, the current British government in its efforts to kick the emergency into the long grass. The political template here is avowedly the original New Deal introduced by Franklin Roosevelt to rescue the US economy from its paralysis after the Great Depression of the early 1930s – and widely opposed at the time, by a whole raft of conservative vested interests, as a lurch into dictatorship.

The rationale for this approach is made very clear in Klein's book from which I have already quoted:

> Environmentalists can't win the emissions-reduction fights on our own. … To win that kind of change, it will take powerful alliances with every area of the progressive coalition. … And to build that kind of coalition, it's got to be about justice: economic justice, racial justice, gender justice, migrant justice, historical justice. Not as an afterthought but as animating principles.

This is written from an environmentalist perspective, but it might equally have come (and in its recent endorsement by sections of the British Labour Party and the US Democrats, very evidently does come) out of a perceived need to reinvigorate a left-of-centre political programme. The nervously unacknowledged premise is that the urban proletariat of classic Marxian theory, not to mention its better-heeled social-democratic successors, has gone irretrievable AWOL in the shopping malls of global capitalism; but here is a replacement, would-be mass constituency for radically amending, if not quite overthrowing, the system which calls forth this rather disparate set of grievances as well as persisting in trashing the planet.

It does not in fact require any particularly tragic kind of realism to suspect that this strategy gives too many hostages to contestability. 'Economic justice' is the most familiar example: does this mean greater equality of income, or of opportunity, or of both? – strikingly different attitudes and policies flowing from any of the possible answers. But corresponding questions can be asked about 'gender justice': does this require only an end to remaining forms of institutional discrimination against women, or is it (additionally) shorthand for superannuating altogether the idea of differences lying deeper than the merely physical in this humanly central domain? Or again, 'migrant justice': a more emphatic commitment to accommodating people fleeing war, tyranny or climate disaster, or the effective abandonment of border controls so that anyone can move anywhere (at least in the West) should they be inclined to try? And it is plain that once such awkward questions are posed, the apparently impressive constituency around 'justice' will be liable to fragment, proving wholly unable to bear the weight of expectation being placed on it. Certainly, advocacy of a Green New Deal on this basis did not noticeably help British Labour in 2019, and was kept resolutely off the 2020 Democratic platform for fear of its helping to re-elect Trump.

But the deeper difficulty with the Green New Deal – and here the relevance of a tragic vision does begin to appear – is suggested by a further remark of Klein's. Nothing about the framework, she claims,

> forces people to choose between caring about the end of the world and the end of the month. The whole point is to design policies that allow us all to care about both, policies that simultaneously lower emissions and lower the economic strain on working people – by making sure that everyone can get a good job in the new economy; that they have access to basic social protections like health care, education and day care; and that green jobs are good, unionised, family-supporting jobs with benefits and vacation time.

It takes a moment or two to make the connection, the registers being so different, but the note of uplift in the latter part of this is reminiscent of nothing so much as William Morris's later nineteenth-century socialist utopia *News from Nowhere*. And much more than Morris, who had his excuses at that stage of history, it is irresponsibly utopian. In this it closely resembles Klein's LEAP Manifesto from her earlier book *No Is Not Enough*, on which her Green New Deal advocacy draws heavily. This is not just a matter of the programme's being offered as a set of win-win practical outcomes – where, for instance, all those whose jobs are lost through abandonment of a carbon-intensive economy will be retrained and re-absorbed into the work of improving energy-efficiency or other carbon-neutral activities. The more fundamental point is that this whole win-win agenda is supposed to be accomplished, by reason and agreement and ultimately through the ballot box, in deference to all the familiar progressive values at once.

Klein specifically cites 'internationalism, human rights, diversity and environmental stewardship' as key values in this context. But, to take only human rights, it is surely clear that conflict between currently recognized rights – to property-ownership, freedom of movement and individual self-determination, most obviously – and the 'rights' of present and future people to a habitable planet, is going to be absolutely central to this transformation. That conflict of that order may well be incapable of resolution on any win-win basis is simply glossed over in presenting the programme; but it is what any tragically informed sense of realism would lead us to expect.

Even more to the point, conflicts as drastic as that are highly unlikely to be grappled with just by reason or argument – they will quickly descend to being fought over. And the example of Gandhian *ahimsa*, so often tacitly in the background here, is actually unhelpful in this context. Gandhi tends to be treated as an unquestionable icon by green activists, but I recall the comment on his activities by George Orwell, who knew British India from inside, that despotic governments – and there are more and less overt forms of despotism – can stand 'moral force' till the cows come home. (Britain left India not principally because of Gandhi, but because it had depleted its military strength and its treasure in defeating Hitler.) In any case, when it comes to seriously defending its interests against a programme of expropriation, which is what both the LEAP Manifesto and the Green New Deal amount to, global capitalism is going to deal far less tolerantly with its opponents than did the Raj, on the whole, with Gandhi; and it will rely on state monopolies of violence to do so. Thus when Klein writes elsewhere that any plausible attempt to reduce global emissions properly 'will mean *forcing* some of the most profitable companies on the planet to forfeit trillions of dollars of future earnings by leaving the vast majority of fossil fuel reserves in the ground' (my italics), and adds: 'Let's take for granted that we want to do [this] democratically and without a bloodbath', the responsibly realistic comment would have to be that we almost certainly won't be so lucky; nor, if we are taking our tragic exposure seriously, can we build the hubristic demand that we should be so lucky into our plans for transformative action as any sort of bottom line.

The Green New Deal, in fact, like all the not-too-dissimilar and long-standing Green platforms which it broadly reproduces, is an appeal for a democratically sanctioned and bureaucratically managed turning of Western civilization upside down and inside out, a caucus-race 'revolution' in which, somehow, everyone emerges with prizes and *no one gets hurt*. To hope for that is to hope for transformation without tragedy, and it should now be clear what is meant by calling this irresponsibly utopian. Such a programme is fundamentally a betrayal – it forsakes the real so that its proponents can feel comfortably righteous, in plentiful company and on the side of history.

It is pertinent here to take note of a sobering observation from the scholar of revolution John Dunn: 'The populace at large does not rebel for fun

… most men, particularly in countries in which massive rebellion is at all probable, rebel as a final gesture of misery, not as an expression of optimism about the future'. Here the relevance of a tragic framing becomes fully apparent. For what Dunn is pointing out is that significant political and social revolutions only happen when a system confronts an explicitly tragic crux – that is, when some central feature of it has become intolerable enough to enough people that the whole system has to be changed *at whatever cost*, in some direction which at least seems to offer escape from the intolerable. Such drastic change will inevitably come at the sacrifice of important values which in an ideal (non-tragic) world people would have wished to continue respecting. That recognition sets a standard for responsibly hopeful usage of the concept of revolution in the context of climate emergency – a standard which on this showing the Green New Deal, with its aspiration to win-wins across the board, fails to meet.

Extinction Rebellion

Does the Extinction Rebellion (XR) movement measure up better in this respect? At first blush it would seem to do so, since an essential part of its self-conception and self-presentation is as a response by increasing numbers of ordinary people of goodwill to a situation recognized as, indeed, intolerable. The co-founder of the movement, Roger Hallam, sets this out very starkly in his pamphlet *Common Sense for the 21st Century*, explicitly and powerfully modelled on Tom Paine's 1776 primer for the American Revolution. It is possible, as the idea becomes more familiar in public discourse, to talk fairly glibly about the climate emergency and its potential for environmental catastrophe; but

> Let's be frank about what 'catastrophe' actually means in this context. We are looking here at the slow and agonising suffering and death of billions of people. A moral analysis might go like this: one recent scientific opinion stated that at 5°C above the pre-industrial mean temperature, we are looking at an ecological system capable of sustaining just one billion people. That means 6–7 billion people will have died within the next generation or two. Even if this figure is wrong by 90%, that means 600 million people face starvation and death in the next 40 years. This is 12 times worse than the death toll (civilians and soldiers) of World War Two.

The XR movement is based squarely on the premise that as more and more people are led by the growing prominence of climate derangement and ecocide to make these appalling connections and decide that enough is simply enough, they will increasingly be prepared to commit themselves personally to rising up against the economic system and in particular the governing arrangements which have brought the world to this unprecedented pass.

XR has too a clear strategy for converting recognition of climate and ecological jeopardy into a source of energy for transformative political change. This involves mounting an intensifying series of acts of non-violent civil disobedience, involving enough participants to disrupt or even partially close down major cities, in particular capital cities as the seats of national governments. Media images of people in large numbers and of all ages getting themselves arrested for obstruction while demanding changes which they patently believe to be in the interests of everyone, will then generate – as well as much greater public familiarity with the issues – widespread sympathetic identification, leading others to join in further mass disruptions. Research shows (it is claimed) that only 3.5 per cent of the population needs to get drawn into these activities before government finds itself faced with the alternatives of either unleashing dangerously provocative repression (risking an explosion of popular anger), or making concessions to the campaign's demands. Those demands are, firstly, that government should tell the truth about the climate emergency; secondly, that it should commit itself to net zero greenhouse gas emissions by 2025; and thirdly, that it should set up a deliberative process based on Citizens' Assemblies with the power to identify and mandate policies for meeting this target. The assumption must then be that something like a Green New Deal framework, as set out towards the end of Hallam's pamphlet, would emerge from this process, although the campaign itself does not seek to second-guess its Assemblies in that regard.

This strategy has the merit of offering campaign goals which at least look achievable. That impression, indeed, was powerfully reinforced when, after the April 2019 series of mass disruptions in central London and elsewhere, both the UK Parliament and (at the time of writing) more than 250 local authorities in Britain appeared to respond to the first demand by actually passing resolutions recognizing the existence of a climate emergency. But that, of course, was the easy part. The demand that the UK achieve carbon neutrality by anything like 2025 is much less reconcilable with even a reformed version of politics or economics as usual. The scale and speed of change required to phase out most of our carbon-dependent activities within five years (a requirement, as the science says clearly, if we are to have even a 50 per cent chance of avoiding climate catastrophe) could only be accommodated by putting in progress a drastic transformation in our entire way of life. Hallam's pamphlet already quoted pulls no punches in cataloguing the inadequacies of reformist approaches to climate change – the whole sorry tale of self-satisfied political and institutional *incuria* which has resulted in the now-looming emergency:

> NGOs, political parties and movements which have brought us through the last thirty years of abject failure – a 60% rise in global CO_2 emissions since 1990 – are now the biggest block to transformation. ... They offer gradualist solutions which they claim will work. It is time to admit that this is false, and a lie.

Mass civil disobedience might therefore seem to be the only means left to drive forward change of the order required in the time available. Is XR then, the genuinely revolutionary movement of which we now stand in such urgent need?

In fact, it has major weaknesses of essentially the same kind as those which beset the Green New Deal. Where that approach places excessive weight on the contestable core idea of 'justice', XR relies much too uncritically on the proposed 'democratic' mechanism of citizen deliberation.

Its model for Citizens' Assemblies is familiar enough from the literature on deliberative democracy and to a limited extent also from recent practice. A group numbering (depending on topic and context) between two dozen and two hundred members of the general public, socio-demographically calibrated to mirror the population at large, is assembled through random selection ('sortition') to address a specific policy question or questions. The group is comprehensively briefed by relevant experts and stakeholders, and professionally facilitated in careful deliberation leading to the production of agreed recommendations. The process is supposed to yield a much stronger representativeness than can be achieved by elected legislative bodies, many of whose members will be people for whom many electors didn't vote. It is assumed that decisions reached by a randomly selected cohort of ordinary citizens deliberating seriously under conditions of maximal information ('exposed to 360° understanding of an issue', as Hallam puts it) would in principle have been reached by any other such random cohort, and therefore that these decisions, since they would have been agreed on by anyone, can reliably be taken to stand in the name of everyone. As a corollary, citizen deliberation is also claimed to be able to take on issues too difficult or potentially divisive to be tackled by elected politicians with their eyes on short-term re-election, exposed to all the pressures of corporate and other influence and operating under 24-hour media scrutiny: its upshots will have a correspondingly greater democratic legitimacy.

These claims for the approach are clearly not to be dismissed out of hand. Well-conducted Citizens' Assemblies might significantly change the political dynamics of many controversial and conflicted issues, or at least help to do so. The trouble is that XR offers them as a kind of panacea – not merely a useful consultative or advisory adjunct, but a mechanism to which legislative authority for dealing with the climate emergency should be swiftly handed over. To this idea there are two serious objections.

The first is pragmatic. Like the Green New Deal's putative coalition around contested notions of justice, remitting the climate emergency to assemblies of this kind simply gives too many hostages to contingency. Even if established politicians could be persuaded to concede real legislative power to such fora – perhaps by giving them some degree of formal input into the deliberative process – the fact remains that cohorts would be selected, however randomly, from an electorate among whom less than three people in every hundred voted for a credible Green agenda at the most recent opportunity (the December 2019 British general election, as of the time of writing). And recall that this is not much less than the 3.5 per cent which XR thinks it can gain its demands by getting actively out on the streets – so its Assemblies would still be reflecting more or less the current attitudinal profile of society. There is indeed now in Britain a much wider public recognition of the danger than there was even a couple of years ago, and XR is due much of the credit for that. But recognition is one thing, and signing up to turning your life-arrangements inside out is another. Even given full information, exhaustive discussion and the best will in the world, it is implausible to suppose that random groups culled from this electorate could rise far enough above interested motives and residual partisanship to take all the hard decisions and commit to all of what will inevitably be seen as the sacrifices, now necessary to retrieve a humanly habitable planet.

The deliberative process, with its juxtaposition of sometimes harshly antagonistic viewpoints under the aegis of respectful rational discussion, can indeed change minds and reconcile even startling differences. But appealing, as XR partisans routinely do, to such successes in other contexts and on other topics is misleading here. The examples which tend to get cited are all of minds changed and decisions reached on limited and containable, albeit controversial, matters – abortion and homosexuality in Eire, immigration in Belgium, flood protection in Poland ... – but there is no precedent at all for dealing in this way with the fundamental, across-the-board economic and social transformation which the climate emergency now demands. And when we talk of 'vested interests' in the current system, we should recognize the most massively vested such interest to be actually the interest which the ordinary unreflective majority has in that system's continuance, operating so powerfully because it constitutes the overmastering interest of the addict in his continuity of supply. 'Finding one's soul in commodities', as Herbert Marcuse once put it – in iPhones, cars, flat-screen TVs and holiday flight-packages, to update his examples – is now the principal source of substitute life-meaning for very large numbers, in the West and North and increasingly worldwide, who (being human) can't live without at least a simulacrum of meaning and purpose, but who are denied access to the realities by the lapse of live religious belief and the destruction through technologico-utilitarian 'progress' of the living humane culture which might have served instead. (Nor can we now be as naïve as Marcuse was in the sixties about the prospects of pop-cultural 'liberation' from this condition – or of pop-cultural anything, except more commodification.) The conscious ego, closed off from the dimension of inner human depth where life realizes itself for the individual as its own unquestionable purpose, tends strongly to identify itself with commodified objects and experiences because the capitalist logic of their restless competitive development offers, as we have already noted, *ersatz* life-direction. The majority now thus addicted will in due course be forced by climate disaster into some very bleak cold-turkey-type experiences, negotiating which will be one of the scariest things about the coming breakdown; but in the meanwhile, no radical voluntary resiling from the fossil-fuel economy can be expected from them, any more than the addict can be expected to reform simply from being told, however vividly, about how she is damaging herself. Correspondingly, citizen deliberation which did actually manage to come up with proposals for transformation sufficient to meet the demands of the emergency would be much less likely to be greeted by the popular majority with "There but for the chance of sortition speak I" than with "Who the hell are this lot telling me what I can and can't do?"

The expectation that such deliberation would shirk the difficult issues is certainly reinforced by the experience of the UK Climate Citizens' Assembly which met from January to May 2020. This was not, to be clear, an attempt to implement the XR template – it was convened at the behest of six House of Commons Select Committees to consider how Britain should meet its recently legislated target of net zero greenhouse gas emissions by 2050. No doubt the publicity given to XR's demands in 2019 helped push the exercise up the agenda, but it had nevertheless a purely advisory function – in the rather weaselly terms adopted by the committees, they would 'aim to use the assembly's results to inform their work in scrutinising government'. Its statistically representative nature also ensured that only slightly under half the Assembly's membership described themselves at the outset as 'very concerned' about climate change. So its results cannot be treated as straightforwardly predictive of what might emerge from an Assembly set up as a consequence of forceful XR pressure, after significant consciousness-raising, operating over a longer period to encourage familiarity of participants both with each other and with the evidence, and with a remit to address a very much more exigent target date for net zero. But they can still perhaps be taken as broadly indicative. And what they indicate, unsurprisingly, is a strong general tendency to appeal to technology and 'better information' as a way of avoiding hard choices. The recommendations, subject to a principle of ensuring *fairness* which apparently took priority over that of *urgency*, include minimizing restrictions on land travel which might affect lifestyles, while limiting growth in air passenger miles (yes, that's limiting *growth*) to between 25 per cent and 50 per cent up to 2050, so that 'freedom and happiness', along with the economy, are not too seriously interfered with. Electric vehicles and aircraft, and synthetic fuels, are to supply the solutions relied on here. In the field of personal consumption, meanwhile, information and incentives rather than regulation are to rule: all changes in diet must be voluntary, and there was explicit majority opposition from participants to changes in taxation or working hours, as well as to any sort of personal carbon rationing. It is painfully evident that this clutch of good intentions fails every test of adequacy which could be seriously proposed. Plainly, if we recall the 'carbon cliff-edge' illustrated in Chapter 1, such measures (even if they did get past 'informing' parliamentary discussion, and were actually implemented) would be quite insufficient to achieve net zero even by mid-century, never mind by the much earlier date which the science shows to be required.

That objection by itself might not be decisive. Attempts to grapple with climate crisis through the existing political system have indeed so lamentably failed, that if we merely thought that citizen deliberation represented our last best shot at doing *something*, we should still be obligated to try it, even while recognizing that there are no guarantees. But as with the Green New Deal, these pragmatic difficulties are based on an objection which runs much deeper.

Hallam's own account makes plain what this objection is. He contends for Citizens' Assemblies on the grounds that they create 'a forum where deliberation and reason will finally be given space to trump the power and corruption of big money'. Unimpeachably democratic, they will allow hitherto-excluded ordinary people to break free from a deeply compromised system dominated by capital and 'neo-liberal elites' which is threatening humanity with extinction. Again, their introduction will channel the growing rage and impatience of these ordinary people at 'unaccountable global elites' who have 'been robbing us for 30 years [and] are now going to take us to our deaths'. And all this represents not just an implausible panacea, but a refusal of the tragic both in diagnosis and in proposed remedy.

Revolutionaries do, of course, need to simplify. Brutal concentration on the shortlist of things that *must* be done, at whatever cost (carbon rationing and the rest, as previously suggested) could only help to keep us engaged with the realities we face. But there is an important difference between doing that and simplifying *out* the genuine tragedy of our plight, which, to repeat, is tragic in the full sense of the term. That plight has arisen, as already noted, because of deep-seated contradictions within the Enlightenment spirit. The human strengths going with rationality also bring with them aspirations to mastery and control which have now jeopardized the biosphere. It follows that attempting to blame everything on elites or corporations, or even on capitalism as a whole, while an understandable reaction, is essentially an exercise in scapegoating. Corporations, for instance, exercise irresponsible power, create deleterious pseudo-needs through advertising and cause ecological havoc in pursuit of short-term financial interests. But they could not do these things, indeed they would not exist in their current forms, had not aspiring billions across the globe (taking their cue from, but no longer confined to, the West) been eager to buy their products and benefit materially from their innovations. Correspondingly, we have brought this disaster upon ourselves not in the main through the traditional vices of pride and hatred, but by the over-indulgence of what are in themselves perfectly creditable passions and desires – for empowerment, for recognition and respect, for general material betterment (that is, for the elimination of squalor, hunger and disease, as well as for lives smoothed and facilitated by 'consumer goods'). And in this tragedy, almost everyone – including almost anyone who might be randomly selected by sortition from among 'ordinary people' in a country like Britain – is thoroughly implicated.

The tragic trap which the Enlightenment set for itself from the outset has in fact now closed upon us. But the citizen-deliberative panacea proposed by XR represents a last-ditch attempt to refuse recognition of this fact, by reasserting a version of the central Enlightenment delusion which parallels that informing the Green New Deal – the delusion that rational humanity can still have it all: in particular, that our key social-political values, justice *and* democracy *and* liberty *and* technological mastery *and* universal material well-being on a habitable planet, can all be achieved together. The peril here is not just failure to see that these values have been brought by ecological reality into increasingly intractable conflict. More fundamentally, it is that we thereby nullify the only kind of hope that could now save us, the deep-seated life-hope which is capable of recasting the empirical odds. Releasing the power of such hope within us depends, as I have argued, on acknowledging our tragic nature and situation, our liability always to be caught up in real and painful dilemmas beyond the reach of reason. Only thus can we repossess that sense of ourselves as whole, embedded, challenged and struggling but still potentially creative beings, out of which the force of life-hope perpetually springs. But any approach which still yearns, however tacitly, for the utopian fulfilment of all our accustomed values, is not going to be up to tragic realism, nor therefore to the revolutionary change in which that hope must now be invested.

Both the Green New Deal and XR are, in other words, still fundamentally reformist enterprises, because they are both structurally attached to the full spectrum of Enlightenment values within which they have been conceived. What distinguishes the revolutionary from the reformer is readiness to recast and reinvent such values *as well as* the institutions and practices reflecting them – and this readiness has always a tragic resonance. Hope invested in the possibilities of transformation, that is, will indeed take us *towards* revolution.

Only through the acceptance of our situation as tragic, however, will it take us all the way.

But if neither of these approaches will do, what would jettisoning reformist inhibition actually involve, and what kind of genuinely revolutionary force would it need to release? That is the question for the next chapter.

8

The New Revolutionary Dynamic

For all that I have been saying in Chapter 7, XR is evidently the nearest thing to a climate-driven revolution yet to have emerged in the advanced West, where the emergency must be decisively confronted if there is to be any future for humanity. It draws, as we have seen, on genuine energies of recoil, rejection and revolt. The importance of those energies cannot be overemphasized, but nor must their nature and dynamic be misconstrued.

We can bring out what is at stake here by reflecting on a characterization of Rebellion which might strike one, prima facie, as peculiarly insightful:

> This opposition, which does not have the traditional class basis … is at the same time a political, instinctual and moral rebellion. … A strong revulsion against traditional politics prevails: against that whole network of parties, committees and pressure groups on all levels; against working within this network and with its methods. This entire sphere and atmosphere, with all its power, is invalidated; nothing that any of these politicians, representatives or candidates declares is of any relevance to the rebels; they cannot take it seriously, although they know very well that it may mean to them … going to jail, losing a job. They are not professional martyrs. … But for them this is not a question of choice; the protest and refusal are parts of their metabolism.

This emphasis on the 'metabolic' combination of 'political, instinctual and moral' motivational elements – the gut and the conscience working together – tells us something very important about XR's genesis. But the conjunction becomes still more revealing when we register that we have here not, in fact, a description of the current situation at all, but part of the neo-Marxist Herbert Marcuse's enthusiastic welcome for the student movement of the later 1960s. That is, or should be, uncomfortable as well as revealing, because we know what happened that time around. Those student protests, described unsympathetically but not unfairly by Alasdair MacIntyre as 'more like a new version of the children's crusade than a revolutionary movement', were an expression of typical adolescent discontent which happened to coincide with widening access to the protected space of tertiary education, increasing availability of contraception, the beginnings of the televised 'media cycle' and an easy target for moral outrage in the Vietnam War. The 'new sensibility' which Marcuse so precipitately welcomed, was in fact a licence for immaturity to proclaim itself as something else, and subsided quickly from outrage into complacency, eventuating in no revolution at all, but rather a powerful reinforcement of the most trivializing and alienating manifestation of capitalism, the commercialized pop 'culture'. And the point of that cautionary tale, of course, is that we simply have no time now for another such self-gratifying 'rebellion'. Something has to be radically and vitally different this time from the retrospectively fragile afflatus trumpeted by Marcuse, if our present recoil from a lethal system is to generate the hard implacability required for genuinely hopeful transformation. And again, we shall find that in exploring for the necessary difference, it will help to call on a tragic understanding of our condition.

Rebellion, morality and the intolerable

What needs to change, in fact, for this present movement of rejection and transformation to have any chance of succeeding, is what is expressed in that linking of the instinctual and the *moral*. Here we pick up again a strand of argument from earlier in the book. In Chapter 4, I argued that morality offers a misleading template for the kind of responsibility which we need to deploy against climate crisis. The corollary for the present chapter is that neither is moral outrage adequate to supply the motivation for revolutionary action appropriate to those real responsibilities.

Marcuse, author of *Eros and Civilisation*, himself thought that a late capitalism which repressed libidinal life-instincts in order to motivate unnecessary production for meeting manufactured 'needs', and which simultaneously offended moral sensibilities (among the young in particular), would thereby give rise to a new kind of urgency of resistance – one which might substitute for the classic Marxist 'immiseration of the proletariat' as a driver for revolution. Conscious righteousness riding on instinctual revolt against repression would emerge as an organismic need, in face of which the constrictions and embedded evils of the system would present themselves as unbearable – perhaps to the degree required, as indicated by Dunn, for rebellion to turn into revolution.

A similar emphasis on morality as motivating force, it will also have been noted, reappears (though without any of the Freudian subtext) in Hallam's calculation of the looming death toll from oncoming climate disaster, quoted in the previous chapter. There, what is to confront us as intolerable in that desperately energizing way, prompting a quasi-instinctual reflex of revulsion, is the sheer scale of what is threatened: billions of people, at best guess, starving over the coming decades as a direct consequence of global mean temperature rise and associated ecological collapse. And this is not just an appalling prospect in itself: it turns readily into the object of moral outrage when we consider that the carbon-profligate world community has full access to a scientific knowledge-base which both warrants such a prediction, and also indicates the kinds of measure needed to avert or at least significantly mitigate its fulfilment.

It is important to recognize, recalling what was being said in Chapter 4, that such outrage can be embraced even in the absence of any genuine nexus of moral obligation linking ourselves and these imperilled billions. Paradoxically, indeed, that absence might make things easier for the outraged. And this points to the root of the trouble here.

Intolerable is a word which gets rather lightly used in a culture habituated to socially mediated outcry against whatever some interest group or specialist community has most recently deemed offensive. But what is sufficiently intolerable to generate revolutionary energy, must be what in some vital sense *blocks life*. Physical deprivation of the immiserating kind (ultimately, denial of the worker's basic life-needs through the relentless logic of surplus-value expropriation, on the Marxian model) clearly does this – life is blocked in the most crude and obvious way for people driven below a minimum subsistence level. The difficult thing to have to recognize, however, in an essentially secular culture where so much weight is placed on moral concerns as giving shape and seriousness to human affairs, is that these concerns by themselves simply do not confront us in that way. The state of finding something morally repugnant can certainly act as a powerful motivator. But it does not correspond even at its most disturbing to immiseration in the sense required to trigger a revolutionary response.

Importantly, this is not to diminish the force or significance of genuine moral engagement. Such engagement can demand kinds of courage without which humans could not flourish, frequently including the courage to recognize what is morally repugnant as demanding protest or remedial action, however uncomfortable, difficult or frightening one finds this demand to be. What is at issue here, however, is the relation of such a challenge to the life-possibilities of the individual responding agent – in particular, the possibility of being brought to such an intolerable stop, of finding one's life-drive blocked in such vital ways, that removal of the blockage is demanded at whatever cost, by a kind of reflex like that of gasping for breath when suffocating. The exercise of moral concern is actually so far from being an experience of that kind, that it can often be felt as the free flow of life-energy into action – emphasizing our shared humanity, assembling us with others around a common purpose, inspiring acts of commitment and altruism which build self-worth on the basis of cooperative endeavour. No claim is here implied that such experienced positives are necessarily the underlying *point* of morality, so that any concern for others is always directed 'ultimately' towards one's own psychological well-being. But that attempt to 'unmask' the process is a crude vulgarization of the really difficult truth about how life works in this domain – the truth that activity to oppose perceived moral evils goes readily *with the grain* of life for a radically social primate consciousness, insofar as it is inherently goal-directed, constructive and collaborative.

That truth is all the more difficult because (inevitably, where life has become self-conscious) pathologies of various kinds also lurk constantly around these processes. One does not have to go all the way with Nietzsche, seeing values like equality, compassion and justice as a comforting conspiracy of resentment mounted by the weak against the strong, to recognize that more frequently than we should like to admit, moral engagement has its unadmitted rationale in satisfactions covertly pursued by the subject of the concern, rather than in the deplorable condition of its object. This is not just a matter of the hectic thrill which so evidently underlies so much moralizing on social media, detached as that is from the restraints of genuine human engagement – where the thrill is clearly serving as compensation for deranged life, for the breakdown in an alienating society of many natural sources of satisfaction (kinship, place, progeny or creative work). Even in much less distorted contexts we can observe, if we can muster sufficient detachment, the process whereby more and more far-fetched 'rights' and implausible ramifications of 'injustice' are ongoingly invented, essentially in order to provide moral concern with a continuing supply of apparent objects. And this is on top of the even more routine occurrence of what might be called the Mrs Jellyby syndrome (after the character in *Bleak House*), where moral urgency directed at abstract others (distant people, or people in general as 'rights-bearers') serves as a cloak of concealment under cover of which genuine, maybe onerous or tedious life-duties can be neglected. Characteristic of all such pathologies is that in one way or another they represent the deployment of bad faith in order to preserve the ego-self essentially unhindered in its career, albeit proceeding by hidden or at any rate unacknowledged channels. So far from blocking the life of the ego, they offer just such protected routes for it to move happily onward.

Now it should be plain enough from what was observed in Chapter 4 about the tendency of moral responsibility asserted in face of climate crisis to become a more or less well-meaning charade, that its motivating force as climate-related outrage is liable to self-subversion in just this kind of way. Tacit awareness of the lack of genuinely moral obligation in this arena acts as a permanent release-valve for any potentially explosive pressure which might build up, constantly deflecting life-energy away from harshly intractable real obstacles into the reassurances of collective righteousness. The green movement, especially in its would-be political forms, has suffered from versions of this tendency since its inception, as anyone familiar with its history can testify; and a continuing danger for more ambitious movements like XR is their liability to attract in disproportionate numbers people seeking similar comfort in shadow-moral commitment.

But even when real moral obligation is in play, the best that moral outrage can yield is strong motivation towards the reform of identified particular evils, within a framework supplied by trying to maintain all key values intact. This follows from what might be termed the *seamlessness* of morality. A moral commitment will turn irresistibly into a commitment to respect all moral claims, just through the tendency towards maximum generality inherent in moral reasoning. That everybody deserves moral consideration turns, by way of this generality, into their deserving *every* moral consideration – so that someone who claimed to be committed to truth in his relations with others, but uninterested in justice, for example would reveal himself as not genuinely committed at all. This recalls the ancient idea of the unity of the virtues. And by the same token moral dilemmas – the need to sacrifice liberty to security, for instance – are met with as regrettable just insofar as they interfere with seamlessness; they are occasions where one has sadly to fall short of the overall moral ideal, rather than occasions where the structural inability of morality to bear the weight of our complex human nature is exhibited. All this is no doubt why Marx, as a serious revolutionary, remained so ambiguous in his mature writings about the moral case against capitalism. While the terms in which he described the procedures and effects of that system ('embezzlement', 'robbery', 'exploitation' ...) made it clear that his opposition to it was at least in part ethically driven – that he loathed its hostility to human flourishing – he resolutely refused to line up with utopian socialists like Robert Owen who argued and worked against it on the grounds of its injustice. What its ethical bankruptcy demanded, thought Marx, was its complete overthrow, and what would develop the force to overthrow it would not be conscience-driven ameliorative concern, however energetic and well intentioned, within the seamless framework of the virtues, but the proletariat finally rejecting its entailed immiseration out of desperate life-need.

Revolution and immiseration

So it is not just that, as Chapter 4 was arguing, the template of moral obligation does not really capture our responsibilities in relation to biocidal climate-alteration: the displacement of our revulsion into moral outrage actually has the effect of dissipating the force for transformation which that revulsion should bring to bear. While genuine moral concern can obviously be a powerful motivating force – as witness many important reforming figures from history (Wilberforce, Shaftesbury, Howard as well as Owenite socialism), recoil from our present condition has to grip at an *existential* level, has to confront an intolerable blockage of vital functioning, before it can provide enough energy to drive rebellion to the lengths of revolutionary change. Once it does so, however, the situation is revealed as tragic, because life-blockage then has to be removed at *any* cost, and in that process major values will come inevitably into conflict. A national liberation movement will find its need for political freedom in conflict with the injunction not to harm the innocent, a move to economic communism will find fair distribution in conflict with established legal rights which must simply be overridden – and a movement to save humanity from unprecedented global peril which it has brought on itself through neglect of the laws of ecology will have to restrict large tracts of what had long been celebrated as human liberty. Moral motivation, in both good and bad faith, doesn't just go too readily with the grain of life to bring us to any such revolutionary crux: in the seamless nature of morality itself, it works against recognition of such tragic necessities and in favour of an always tempting utopianism. But utopianism, as we have seen, disables genuinely transformative hope.

Hence a movement like XR, if it is really going to spur us to save ourselves in the time we have left, needs very urgently to move beyond morality as the main ground of its appeal. Inheriting as it does the mantle of forty years' environmentalism, concern about the effects of current behaviour on future generations and the prospects for other living species will of course remain at its heart. But how, addressing this future out of a still-habitable present and for people whose eyes are mostly, and understandably, on their immediate circumstances, can it bring enough of them up against the presently, 'metabolically' and literally *intolerable* to provoke a social explosion having the necessary force?

We should not expect to answer this question in the same terms as for previous revolutions. Part of recognizing our current situation as unprecedented is understanding how the parameters of the dynamically intolerable may well be strikingly different from anything which has served that function in the historical record.

It is already clear that the decisive factor will not be anything at all like the material immiseration, the humanly insupportable workload for less than starvation wages, on which Marx counted for the revolution through which the urban proletariat was to inaugurate communism. Something corresponding to that, if not quite as he formulated it, does indeed seem to have been a key factor in at least two of the great historical revolutions, those in Russia and in France. In the former case, mass slaughter on a chaotically disorganized battle front and serious food shortages in the capital cities meant that Lenin's simple slogans of Peace and Bread could strike like hammer blows against the helplessness of Kerensky's Provisional Government. In the case of France, a National Assembly orchestrated largely by lawyers impatient for both fiscal competence and the Rights of Man was pushed from reorganizing a kingdom into running a full-scale social revolution once its activities took the lid off peasant protests against impoverishing seigniorial dues, and also gave rein to urban riots against bread shortages. The revolution which we now need must challenge comparison with those great pivotal upheavals – it too must delegitimize and then overthrow (in the first place, for us, here in Britain, but then very rapidly worldwide) a corrupt and vicious established regime which has long forfeited its right to rule. But it must be launched, in the West, in conditions of material plenty, indeed of gross material superabundance – conditions where, despite the effects even of considerable maldistribution, those lacking the basics of food, clothing and shelter form a tiny, marginal fraction. Only in the most forlorn hopes of the Green New Deal could it be imagined that anything now corresponding to material deprivation would provide the essential impetus for transformation.

That impetus, however, does not *have* to be materially grounded. Marx was notoriously committed to a materialist view of history, but it is nothing more than a defining prejudice of such a view that intolerable deprivation has to be itself material. In the seventeenth-century English Revolution for instance, what played the role of fundamental motivating drive was arguably the threat of *spiritual* deprivation. Although rebellion was triggered by the Long Parliament's legislative and fiscal disagreements with the King, the decisive fighting energy of the Cromwellian armies came from the force of nonconformism breaking out of attempted religious bondage – the direct, personally exposed and passionate sensibility of a Bunyan recoiling from the insufferable constrictions of Laudian ritual and formalism. Again, the American Revolution was hardly driven by economic considerations at all (the taxation issue having been fundamentally a pretext), but rather by the insupportability of arrogantly reasserted colonial leading-strings for a set of communities whose defining sense of their own identity had come to celebrate pioneering self-reliance and self-determination. And yet, in both these cases, what burst through established bonds were forces powerful enough to turn a whole society upside down and re-found it in a new form.

As these last examples remind us, and as Marx (or at any rate the later Marx) was inclined to ignore, humans are in fact the only kind of animal which lives naturally in the realm of values and significances as well as that of material forces. Human life can be blocked sufficiently to provoke transformative outbreaks as readily when circumstances mean that for one reason or another the *spirit cannot breathe*, as when the stomach is clemmed with imposed material want. But it should also be clear that no more readily than material immiseration do non-material revolutionary precedents from these particular historical examples transfer easily to our present plight. This is (despite freaks and throwbacks) an overwhelmingly secular age in the West where it most matters, and nothing remotely like the religious energy of Puritanism is now available to us for resisting ecocide. Nor are the Rights of Man or the clarion call of Liberty to the point, except as indications of how deeply we are now entangled in mistaken relations between humanity and its planetary context. So the question becomes: where in current conditions can we point incontrovertibly to an *equivalent* suffocation of the spirit, affecting people in such a way that their struggle to fight free of it might have the chance of making a revolutionary difference?

The answer to that question turns on the absolute need which each of us has for his or her life to be experienced as meaningful, and on the conceptual connection between that need and an absolute rejection of the ecocidal, or more strictly biocidal, tendencies of present civilization. Put simply, what is needed to power the overthrow of that civilization's destructive governing arrangements is the recognition that they radically attack the possibility of life-meaning, without which human beings are intolerably impoverished as surely as without food they starve. This unprecedented revolution will spring out of recoil from a kind of immiseration which might be called, as already noted, neither material nor strictly spiritual, but *existential.*

Existential immiseration

In his prescient 2006 book *Heat,* George Monbiot claims that our use of fossil fuels is a Faustian pact. Faustus was the legendary medieval German scholar and philosopher who sold his soul to the Devil for twenty-four years of magical powers. According to Marlowe's play of 1590, he thought he was getting a good deal, because at the outset anyway he dismissed talk of an eternity of damnation as 'mere old wives' tales'. Once endowed with supernatural abilities, however, he used them to lead himself a dance of initially exciting but thereafter increasingly trivial self-indulgence – by the end of which he thoroughly repented of the deal, though the Devil didn't. Monbiot suggests that this closely parallels what humanity seems now to be doing: using fossil fuels to give ourselves a brief lease of quasi-magical technological near-omnipotence, at the price of a future world which (if global average temperatures rise by the 6°C or more which we are now inviting) will indeed be a fair imitation of Hell. And the parallel extends to our either not really believing, as we enjoy ourselves, that these consequences will kick in; or persisting in simply turning our eyes away from them – or perhaps, as in Goethe's later (and characteristically Enlightenment) twist to the Faust story, thinking that we can rely on our technological ingenuity to mitigate the worst of its own adverse effects, and so always keep one jump ahead of damnation.

At first blush this Faustian comparison seems like a persuasive metaphor for the peril of our situation – even if a slightly self-flattering one: a daring high-stakes gamble on how far humanity can push the frontiers of the achievable at least has something adventurous about it. But on further reflection, a crucial disanalogy begins to surface. Faustus' deal attracted him as appearing to involve a genuine *quid pro quo*: the reality of extraordinary powers in the present, in exchange for future torment if that should indeed turn out to be more than an old wives' tale. But a deal which even potentially jeopardized the continuing integrity of the Earth's living systems in conscious and deliberate exchange for some supposedly compensating present benefit would be no deal at all, because its terms must annihilate the value of any such benefit which we might have imagined ourselves acquiring, emptying it of meaning in the moment of its acquisition. It would be as attractive an arrangement as remortgaging your house in order to be able to afford the wherewithal for blowing it up. We cannot cut an explicitly Faustian deal here because to do so would destroy the intrinsic mattering on which life-meaning, and therefore any humanly habitable present, depends.

If you doubt this, test it in your own case. Think of what really and centrally matters in your life, and then ask if it could go on mattering to you in that way if your going on enjoying it had been part of a trade-off in which you quite explicitly acquiesced in the human-driven destruction of the biosphere by some undefined point in the medium-term future, once you were safely dead. You will find, if you are honest with yourself about this, that no retention of real significance or purpose is possible under those conditions. But life-meaning is, on any account, a profound human need: indeed, the neurologist, psychiatrist and death-camp survivor Viktor Frankl argued forcefully that it is *the* most profound, the primary human motivation. Explicitly to recognize such a Faustian 'pact' as the rationale for what we are doing with fossil fuels and the climate must therefore at the same moment be to reject its terms categorically. And to be caught up against one's will in a system which is evidently operating on those terms is to be subject to a wholly intolerable form of oppression.

What is the necessary relation between life-meaning and the ongoing integrity of the biosphere, on which I found these claims? In brief: meaning is given by what we can't step back from and see as merely conditioned in our lives; but the only thing from which we finally can't step back in this way is *life itself*, experienced as an active principle of energy and spontaneity within us; and that foundational recognition cannot consort with our knowingly and wilfully travestying the basis of living energy and spontaneity in the wider world. And here must again be emphasized the close connection with what was being argued in the latter part of Chapter 4. For I might put this point in different words by saying: life can now only be robustly meaningful for someone consciously and deliberately life-responsible in the way there described. The foundational recognition is the dawning of that kind of responsibility.

How do I make that out? Human lives typically acquire meaning through concerns which give direction or simply a kind of vigorous sufficiency-in-itself to the ordinary business of going on. Meaning comes not from anything transcending normal, central human satisfactions – those of friendship, food, art, sex, parenthood, work, achievement, exercise … – but from experiencing those satisfactions (or enough of them) from day to day as not empty. This in turn requires that some things matter to us in a way which we can all recognize – that is, they matter *intrinsically*: their mattering depends just on their being what they are and not on their relation to something else which is supposed in turn to give them point. But creating and maintaining intrinsic significance in that way, which we have to do in order to live meaningfully as the kind of creatures we are, is always tending to be undermined by the very kind of creature we are: that is, the only creature which can step back questioningly from itself – from its desires as well as its beliefs, its potential deeds as well as its provisional understandings, and its apparent core commitments insofar as they call it to action. In relation to life-meaning, the distancing reflex which goes with human consciousness takes the form of asking, "So what?" It belongs to being a self, that as soon as one thinks of oneself, one is standing back in consciousness from that *of* which one is conscious. The particular interests and claims which give point to what I do or attempt are then seen to arise from the set of relations and activities into which my life-conditions have brought the person who I have become. But so what if these things matter to me as just that person? – since I needn't have been. Everything that might matter in the intrinsic way which grounds life-meaning then only 'matters' in quotes, as it were – and about what only matters in quotes, the "So what?" question can always ask itself again – does its 'mattering' really matter?

And yet, of course, we do constantly find reliable meaningfulness in our lives, otherwise we could not go on finding the world humanly habitable in the way that most of us, for much of the time, succeed in doing. This we achieve through halting the reiterated "So what?" by bringing it up against something fundamental of which we are aware all along as not open to question in that way. The drive *to act* – to grow, develop, strive or move in one direction rather than another – is what fundamentally characterizes *life as such*. For human consciousness, that drive presents itself as an imperative: go with the life-impulse of spontaneous self-realization at the core of your being. And from this impulse one finally cannot step back in any kind of questioning, because it is presupposed in all questioning – it expresses itself in any activity whatsoever which the conscious subject undertakes. The meaning-giving force of what matters to us intrinsically is thus grounded in the deep responsibility to which Lawrence was pointing to when he said, as I have already noted in Chapter 4: 'Resolve to abide by your own deepest prompting. … Try to find your deepest issue, in every confusion, and abide by that.' One's ability to find life habitable at crucial junctures where its meaningfulness is seriously at stake, rather than just taken for granted in the usual diurnal way, has to rely on a sense of oneself as ultimately a subjective force of living nature, with the final unquestionableness which that involves. What most deeply matters for me reflects not what I find myself happening to will, such that I could have willed otherwise or on a different pattern and still have been the same self, but what nothing could have failed to will and still have been *this* living subjectivity, following this particular clue of spontaneous exfoliation. A living consciousness, in other words, finally cannot 'step back' in reflection from *life itself* – experienced from within as individual perspectival energy, ceaselessly emerging and pressing forward through and as *me*, and also intuitively recognized as an active principle which we share with everything alive.

And this is where the conceptual connection with our stance towards this wider realm of life, the self's self-reconstitution as a centre of life-responsibility, becomes unavoidable. The life-energy thus sensed at one's individual core is unquestionable in the way that I have indicated. This fundamental immunity from question is what we might have captured, in an earlier terminology, by recognizing the life-impulse as sacred. Meaningfulness is only to be found in the kind of living deference to the vital force at the root of all formed personhood which that term suggests. But the unchallengeable vitality which it acknowledges clearly resides in *life as such*, life-energy as it flows impersonally through everything living, rather than just as it manifests itself in me. Life thought of in that way is not distinguishable from the great principle of self-maintenance and self-regeneration which keeps the whole biosphere in being. And to recognize that principle as in the same sense sacred and an object of deference, while participating knowingly in a civilization whose overwhelming drive is towards depleting and destabilizing the whole wider life, is then to be involved in a radical and utterly stifling contradiction of feeling. For the sacred, in any sense, is precisely *not* that of which we wilfully jeopardize the richness, diversity and integrity for any reason, still less for the sake of hectic mobility and trivial consumerism in a plastic-and-concrete wilderness. That stark intuited dissonance would drain meaning and point from whatever was bought at the price of such participation, for anyone thoughtfully conscious of paying that price. A Faustian bargain would betray us, literally intolerably, in the act of being deliberately struck.

To put this conceptual connection another way: someone who sees and feels in full, honest awareness what humanity is now doing to life on Earth has no escape into the immediate concerns of private or professional existence. Nor is this finally because of any moral claim – from future humans, other creatures or the biosphere itself – but because such private concerns cannot survive the recognition and still stand firm as offering a personal space of meaningful activity. Thus – and often painfully – does one become a centre of life-responsibility. One cannot shrug and turn away to cultivate one's own garden, since all one's purposes must be intimately and intolerably maimed by any such self-recognized acquiescence. The only way to keep life meaningful, after such knowledge, is actively to reject complicity in the ongoing damage and then, inescapably, to fight against the social and political systems which are causing it.

That might sound as if existential immiseration could be fairly easily palliated by turning to action – so easily, indeed, as scarcely to count as any kind of misery. But matters here are much less straightforward than that, and not just because some degree of complicity in biocide solicits us everywhere in contemporary life. It is true that 'fighting the system' while remaining in large measure signed up to it is always an option, and can continue to serve as such even while part of one's mind recognizes the charade in which one is engaging. Quite apart from these permanent temptations of bad faith, however, to be bound on the wheel of action by recoil from meaninglessness can itself be a form of the immiseration from which it recoils. The most wholehearted activity undertaken in order to hold at bay a perceived threat of this existential order is always still under threat – nothing so haunted by this kind of emptiness, the meaning-void, as the conscious and deliberate effort to fill it. Here as in other psychological domains, resistance can figure not as an escape but as a subtler form of subjection. Freedom comes only, if it comes at all, from losing oneself in full commitment to extirpate a biocidal system for the impersonal sake of life itself: that is why the instinctive drive out of this immiseration generates the dynamic of a wholly unprecedented revolution. And here again we must appeal to a readiness to acknowledge tragedy: for this dynamic has to transform the system at whatever cost, not only to settled habits, expectations and social arrangements but also to those who must overturn them. As Yeats famously observed in the context of a different revolutionary cause,

> Hearts with one purpose alone
> Through summer and winter, seem
> Enchanted to a stone
> To trouble the living stream

– a truth which embodies an especially sharp irony for those whose full awareness of the climate emergency means that they must accept such a sacrifice of their own life-potentialities to the interests of life at large. Tragically, there is no way for them to get wholly clear of immiseration until the world has been transformed.

If this is right, then facing up to the climate emergency must make a uniquely exacting demand. From the premise that only some form of immiseration could suffice to motivate genuine revolution, we seem to have reached the conclusion that anyone who thinks clearly about the issues now confronting us must become, however reluctantly, a thoroughgoing revolutionary, since to biocide honestly acknowledged there is no other habitable response. That might seem at first blush so extreme a result as to call the whole argument into question. But actually, how could things be otherwise than all-or-nothing here? Greta Thunberg has pointed out with her characteristic terrifying clarity that 'If the emissions have to stop, then we must stop the emissions … that is black and white. There are no grey areas when it comes to survival.' And if the opening-up of a void of emptiness at the heart of human life is a wholly new kind of driving force for action, it arises in a wholly new kind of existential crux for human beings, who have never previously during the history of life on Earth been in a position to jeopardize the entire biosphere, still less to understand themselves as on the way to doing so.

Immiseration for whom?

At this point, however, we reach a difficulty which more sceptical readers will for some while have been anticipating. For of course, those who do so understand themselves, though there are now more of them than ever before, remain a very small minority of any population as things are – and those to whom participation in a biocidal system has become literally intolerable are going to be a still smaller one. To launch a revolution, though, don't we need some kind of *mass* immiseration?

If we do, then our prospects under these wholly novel circumstances are bleak indeed. But before drawing that desperate conclusion, we should give some thought to the *kind* of small minority which the very idea of existential immiseration presupposes. Those of us who have been trying for some time to get the climate emergency taken seriously have tended, understandably enough, to concentrate on exhibiting the full spectrum of dangers which people need to recognize, and have correspondingly neglected to think much about the spectrum of people capable of full recognition. For instance, in the Introduction to a recent book, my co-authors and I aspire to catch 'a moment at which humanity summons enough courage to look our morbid present and our probably-awful future in the face. Only by doing so … can we rise up to meet our real situation'. But if this is the hope that by making the facts available and arguing persuasively, we could get *everyone* to confront these dangerous prospects before it is too late, it is surely a completely forlorn one. Rather, as I have come since writing that to see much more clearly, a tragic realism must compel us to hope that an informed and active minority will summon its courage *on behalf of humanity* and take the necessary responsibility. This vanguard minority must be seen as the leading edge of consciousness and intelligence dealing with the demands of an utterly unprecedented situation, and hope for transformative change must be invested in its coming to understand itself in this way and acting on that understanding. What, then, are likely to be the characteristics which qualify it for so doing?

The minority whom the threat of existential immiseration could drive to transformative action is a minority because to register that threat and respond appropriately requires intelligence, imagination, reflectiveness, honesty and courage. Luckily, none of these qualities is in itself all that rare, but even the most resolute egalitarian must admit that their combination in full measure is far from common. And here is already a very important suggestion: that the group about whom we are thinking may be in some sense *necessarily* a small minority. This possibility becomes even clearer when we explore its qualities in more detail.

Intelligence is required even for baseline recognition that there is a climate emergency at all, since that recognition can only be reached through the interpretation of events and information in the light of mediated expert opinion. Not even a dramatic development like the 2019 Australian bushfires has 'climate change' written visibly on its face. What people see and feel on these occasions is at best weather and its effects; the idea of climate, and then the further idea of its destabilization, require an interpretive understanding and contextualizing of these experiences. Such understanding in the individual does not have to reflect a scientific training, but it does presuppose a cast of mind which recognizes the need to take on board, as well as a wide range of information, the analyses of this information offered by relevant expertise. For thinking both critically and to the point about these analyses, it presupposes too a degree of common-sense statistical alertness, enabling probabilities to be compared and associated consequences to be pragmatically weighed. While we may hope that these abilities will be reasonably widespread among a population with at least a basic level of scientific literacy, a full grasp of climate emergency calls also on rarer capacities of attention to diversity and multiplicity – to the ramifying ecological interconnectedness which lies at the root of every specific climate or biospheric risk. The intelligence needed for recognition of our present plight must be adequate to systemic complexity in a way which has never before been demanded of ordinary thinking people.

It should thus be clear that what is called for here is not just native wit, which is fairly common, but educated general intelligence, which isn't anything like as common as it should be. Nor can that educated intelligence be a matter of basic scientific literacy alone. At least as important is *humane* literacy – in particular, enough familiarity with enough of its manifestations in our inherited culture to appreciate the tragic dimension of life which I have been emphasizing, to be correspondingly suspicious of the tendency to represent all difficulties and challenges as soluble, and to bring the practised eye of irony to bear on the more ambitious claims of value. This is now a very much rarer form of intelligence than that required for a science-facing understanding. An educational system increasingly driven by the supposed needs of the economy and a mass media which cheapens and trivializes whatever it touches have together undermined the central role of humane culture in general intelligence for at least a century. So much of what has been ineffectual about the green movement to date, for instance, is attributable to the lack of such a culture – to a well-intentioned but callow ethical utopianism bred up through several generations in essential ignorance of human history and human nature. Absent this vital component, however, the most up-to-date merely scientific literacy is likely to be very much worse than not intelligent enough for registering the full extent of the dangers we face. Correspondingly, the numbers of those who do so will be limited.

Another such precondition is imagination: not the same thing as intelligence of this latter kind, but closely linked. This is the capacity to make an object or event which is not being directly experienced nevertheless present to the mind in something like its actual sensory impact, emotional relations and bearing on thought. It is an indispensable qualification in this context because the climate emergency is overwhelmingly (though no longer quite exclusively, especially in the global South) a matter of *prospective* disaster and the potential for longer-term catastrophe. Thus the consequences which intelligent understanding indicates that we are storing up for our successors and in some cases our later selves – the temperature extremes, the devastated habitats, the storms, floods and droughts, food shortages, mass migrations and armed conflicts – have to be vividly imagined in all their painful destructiveness to get the more-than-merely intellectual grip on biocide, the real emotional exposure to it which makes the idea of complicity in it so intolerable. Again, the capacity for this, though continuous with native wit, needs to have been educated through a developed familiarity with literature, drama and history if it is to grapple adequately with our new plight. It is no accident that the best presentations of what climate emergency will actually mean are in works of imagination like Cormac McCarthy's *The Road* or John Lanchester's *The Wall*. But novels like these are of course very far from topping the airport bookshop bestseller list.

To intelligence and imagination, qualities whose combination already quite sharply delimits the field, we must add for the minority here in question what I have called reflectiveness – although that word is only the least unsatisfactory one that I could find for what is involved. The quality it betokens is perhaps in part that which has kept readers who have got this far engaged with this book: a sense of the importance of conceptual connections, of the conceptual architecture of any field of activity with which we are dealing. Thus the sense that *hope* had to be internally related both to *realism* and to *possibility*, that these relations were unclear and that getting clearer about them now mattered urgently for tackling the climate crisis, was partly what incited me to write – and I assume that an interest in how much clearer I could in the event get has helped sustain the persistence of those who are still with me. But there is a further dimension to what I mean by reflectiveness, beyond this concern for what might be called the logic of practice. This is alertness to conceptual relations as structuring also one's own being: the kind of inner attention by which you recognize, as in the previous chapter, how the possibility of a meaningful life depends on finding things which matter to you intrinsically, and how such mattering is always liable to be undermined by the human capacity for self-distancing from all one's contingencies.

I am not of course suggesting that the properly motivated minority has to be philosophically trained, on top of being intelligent and imaginative – that would reduce its numbers towards vanishing point. These crucial connections will generally be made subterraneously and may well not be brought to any focus of explicit acknowledgement. That is part of the reason why their upshot tends so readily to be reached for in the more familiar but finally unsatisfactory language of morality and obligation. But they will, I suspect, be operative in anyone who finds our current behaviour and life-arrangements genuinely intolerable, such that these arrangements are felt irresistibly to merit overthrow come what may. I guess that they underlie, for instance, what Roger Hallam says in his pamphlet which I have already cited: that adopting XR's strategy for change and working to help implement it 'will make us happier. … It is clear why we are here in this world and what we need to do with our lives.' The life-purpose which we gain from seeing that anything except full commitment to halting biocide must erode life-purpose offers a robustness which this vanguard minority will need for coping with the responsibilities falling upon it.

Then there is honesty. Here again, this means something more exacting than the ordinary disposition of well-brought-up people to refrain in most circumstances from telling lies. It means the much less common habit of not lying to oneself, through any of the various forms of denial available to help us evade or cushion uncomfortable truths which at the same time we really recognize. In the case of the climate and ecological emergency, these forms range widely. At one end of the spectrum there is endorsement of the sustainable development paradigm (we are not recklessly warming the planet, we are participating in the Paris Plan to stabilize emissions at manageable levels; we are not committing biocide, we are substituting natural with technological capital to preserve benefit flows over time …). From this one might proceed via offsetting one's air miles ("I strongly oppose global warming, and have to explain why at this very important conference in Ulan Bator" – here, of course, COVID-19 is actually helping, in a way which may perhaps persist to some degree), all the way to dutifully recycling (after buying and using) a whole shed-full of consumer garbage. Someone liable to be tempted by such recourses – and there is a long menu of others like them – can be as intelligent, imaginative and reflective as you like, but won't be driven by the threat of the intolerable into revolutionary recoil, because all these shifts are precisely about making the intolerable easier to tolerate, and so letting oneself off the hook even while knowing that it remains firmly in place. At the end of the day they can't salvage life-meaning from continuing complicity in biocide, but the end of the day can be made to seem a pretty long way off by the exercise of doublethink – a practice with which our essentially dishonest civilization prompts us at every turn. It is by their private resistance to it, even more than by their public commitment to bearing honest witness, that the vanguard minority which I am characterizing is distinguished.

And finally there is courage, without which none of these qualities can be brought to bear in decisive action. While it is perhaps conceptually linked to honesty, which must include the courage of unflinching attention, it is sufficiently distinct from intelligence, imagination and reflectiveness for all of them to be compatible with the cowardice (or, at best, the lack of confidence) that hesitates to follow through on them. So even in this unprecedented situation we still depend, as humanity always has, on enough people having the guts to risk themselves for what they perceive to be their responsibilities. But since only a small minority will register the relevant perception as things stand, the relevantly courageous among them will be still fewer.

It appears, then, that the need to launch, and initially at least to drive, this unprecedented kind of revolution through the existential immiseration of a small minority is not to be seen as a regrettable drawback, but as something inherent in the nature of the case. It is, in fact, essential to the dynamic force of such an experience of the intolerable that it should be undergone by people whose combination of qualities must ensure to them, in any imaginable human world, minority status.

And yet everything we know about what is happening tells us that *this minority must come to power within twenty years* if catastrophe is to be averted. The whole foregoing book has been a prelude to asking whether, and if so how, that can be realistically hoped for. To those questions, therefore, we must finally turn.

9

The Vanguard of Hope

It may be helpful to begin by linking the argument being made in the previous chapter back to the book's opening. The same life within us which we there saw insisting on itself as undaunted hope for indefinite human flourishing, and thereby recalling us to a deeper realism than the merely empirical, we now see insisting also that we defer to our embeddedness in life at large, on pain of robbing individual existence of any chance of purpose or meaning. Hope and purpose are conceptually distinct. (One might hope for something without thinking that it really mattered, and equally one might hold firm to some cherished purpose although one had got beyond either despair or hope.) But experientially, they are closely intertwined – hope very often giving or strengthening purpose and purpose calling on, or calling forth, the hope which provides its necessary energy. Both are forms taken in human consciousness by the forward-pressing force of life. Hence, the kind of hope which we must now pursue, the transformative possibilities which it must lead us to canvass, the necessarily tragic framing of those possibilities and the absolute recoil from biocide which must drive them to take urgently revolutionary shape, all flow together from the deepest source of power that we know. That power is the strength of life itself, pitted against the gravest threat which life on Earth has faced for millennia, and realizing itself through the human species which represents, as we should never cease reminding ourselves, not only the source of that threat but also the agent which might still (just possibly) overcome it.

Agency, however, lies as we have seen initially with a very small minority. This chapter focuses on that vital sector. Given its necessary qualifications as we have identified them, what is its likely provenance, and what opportunities go with that? By what warrant could it seize those opportunities to pursue full power? What kind of programme of activity might actually take it towards power? And finally, what role would need to be played in this process by a serious green-political party, if we had one?

I must repeat here with yet more emphasis the point made at the outset of Chapter 7: the answers which I offer to these questions are exploratory, and only not more tentatively expressed because I want to spark vigorous debate.

Where *is* this minority?

It is evident that those who combine the qualities explored in the last section of Chapter 8 do not form a class, in the old Marxian sense of a sector of the population defined by its relations to the means of production. The crucial relations for this new revolution are not going to hold between ownership of capital and the sources of labour power, but between ecologically destructive production (and consumption) as such and those with the capacity to recognize it for what it is and do something about it. Hence, the vanguard minority exists wherever that combination of awareness and ability occurs: anyone reading this will, for instance, almost certainly belong to it. But there is of course rather more to be said about where across the social fabric such people are indeed liable to be found.

In the first place, some very general characteristics will unite them which do tend to associate loosely with economic status. Most straightforwardly, the need for a rounded education to foster the kind of intelligence discussed previously will mean that members of the minority will simply tend to have gone to better schools, and probably also through some form of higher education, and such an education certainly correlates with middle-class origins. More subtly, a threat to life-meaning from complicity in biocide will be felt most strongly by people who already experience their own lives as generally meaningful, and a powerful source of this is *vocation* – commitment to a work role which has a clear intrinsic value and also employs many of their human faculties, rather than being routine, repetitive or mechanical. Again, such work is often though not exclusively associated with broadly middle-class origins and characteristics.

There are also general characteristics of the vanguard minority which do not align with economic or vocational status at all. Crudely, these come down to 'skin in the game': people concerned for children or grandchildren whose lives will be directly and savagely impacted by a biocidal future will gravitate towards this minority, as of course will those younger people themselves, insofar as they are already old enough to recognize their situation for what it is. The voice of the school strikers is especially powerful and revealing here, emphasizing as it does that their education is effectively being robbed of its point by failure to address the climate emergency. This is much more than a protest against lack of job prospects in a collapsing fossil-fuel economy; rather it reaches into the existential dimension which we have been discussing. In a biocidal system, any goal-directed activity of self-development, any attempt to possess an inherited culture or accumulated knowledge base, becomes strictly meaningless. (Again, the school strikers' inspiration Greta Thunberg is a key witness here: 'We need to focus every inch of our being on climate change, because if we fail to do so, then all our achievements and progress have been for nothing.')

Within that broad framework, we can identify (albeit fairly roughly) what the occupational profile of the potential vanguard minority would look like, in terms of where the qualifications for belonging to it explored in the last section could be expected to cluster. That this points to a *potential* minority is, however, important to stress. Just because it is not pre-shaped by its economic or social relations, the vanguard of this revolution will need to be deliberately assembled, to be called into being as a self-recognized cohort from among all those who share in the relevant qualities and incentives. That process is what this book aspires to serve, through whatever persuasive power it wields and any discussions which it succeeds in provoking: I was not engaged in flattery when I said a paragraph or two back that readers of these words were likely to be members already of the constituency on which a frightening responsibility now devolves.

The kind of rounded practical intelligence called for, then, is going to characterize a wide variety of professionals – legal, official, administrative and managerial – as well as some officers in the services and anyone in any business enterprise which requires a degree of conceptual agility along with practical acumen. The necessary imaginative capacity will tend to predominate in creative people generally – writers, artists, craft-workers, actors, designers, some engineers and also people involved with the media to the extent that they have resisted selling out to mass-communication values. A disposition to the appropriate order of reflectiveness is likely to be found among those involved with education at all levels (including, as noted, those on the receiving end of it), and perhaps strongly also among caring professionals, those engaged in charity-sector work, and those whose work involves direct engagement with the natural world – all broad categories of people whose vocation is in one way or another the fostering of life.

It is probably best to go no further down the list of qualities than this – it would be very tendentious to try to identify groups or occupations characterized by honesty (if perhaps less so to suggest where its opposite might predominate), and courage can be called for at times by pretty well any occupation. But what even this very outline sketch makes clear is from how broad a spectrum of society the potential vanguard could be recruited. And by the same token, it highlights an absolutely crucial feature. Despite not constituting an economic class, despite not holding crude economic power as such, the range of people whom we have been identifying are those on whom the fossil-fuel state, at any rate in its liberal-democratic form, is utterly dependent. (Let the term 'fossil-fuel state' stand for the moment as a convenient shorthand – its reference is clear enough for present purposes, although we shall need to define it in more detail later on.) These people do not rule the state (though they overlap considerably with the political establishment which does), but without their compliance and implicit sanction, no one can rule. That might not be the case in some populist dictatorship, but in the kind of Western society where the new revolution must happen first if it is to happen at all, it is surely evident that if enough people from all those groups and occupations withdrew their acquiescence and cooperation in a coordinated way for even a week, the state and economy would cease to function.

If the exposure of this putative vanguard minority to existential immiseration by its characteristic qualities explains its revolutionary potential, the firm grip with which, if it came to think and act as a concerted force, it could lay hold on the levers of legitimacy and efficacy of the fossil-fuel state, perhaps indicates its most plausible route to power.

Three bugbears

But if it could take such a route, should it? The nervousness underlying that question may well have burgeoned rapidly among readers of the last section, because three apparently intimidating bugbears immediately loom up. They are the linked objections that for a conscious and coordinated vanguard minority to act along these lines would be undemocratic, elitist and even (if, after subverting the fossil-fuel state, it then took and held power on its own account) incipiently fascist.

The thing to do with bugbears is to be rid of them. We can rid ourselves of these by seeing that objecting to such a pursuit and seizure of power as undemocratic is woefully out of time; as fascist, historically illiterate; and as elitist, properly understood, no objection.

'Undemocratic'

That any response to the climate emergency must preserve democratic norms and values is an assumption running in the bloodstream of the existing green movement, and announced deference to this assumption serves as a shibboleth for inclusion in any respectable discussion of the issues. But those issues are too urgent for considerations of respectability. What if we cannot now retain these norms and values *and* drive through the swift and drastic transformation needed to avert climate catastrophe? This would certainly be a tragic situation, since the claims of democracy speak to aspects of the human condition (the equality of rational selves) with which we must go on reckoning, and moral imperatives (the demands of universalizability) which exert real force. That doesn't mean, however, that it wouldn't be *completely clear* which of those two options, coup or catastrophe, we should go for – since unless catastrophe is averted, there will soon enough be no human condition or arena of moral action to worry about. One important upshot of a recovered tragic view of life would be recognition that this bitter choice has now for some while been confronting us.

As to democracy in current Western practice, indeed, one might simply (and to my own mind, quite sufficiently) say that a political system which even temporarily bestows major leadership roles on persons like Trump or Johnson is very evidently broken, and therefore to entrust it with the future habitability of the planet would be near-criminal folly. Know-nothing populism is not 'democracy's ugly sister', it is what democracy herself is reduced to after a course firstly of the gutter press, then of television and finally of Twitter as prime political fora. But for any reader who does not find this already a conclusive argument, the following considerations should be weighed.

The insistence that we can only save ourselves democratically is out of time. Firstly, and most evidently, democracy in its Western-liberal form has had forty years to address the gathering climate and ecological crisis, and has failed dismally to do so. It is true that over that period these issues have gained a place on, and even risen notionally quite high on, political agendas. But the results of this process when compared with the dangers which it is meant to be addressing can only be called nugatory. Democratic states have sustainability programmes which they treat as dispensable when economic difficulties loom, targets written into international agreements supported by these states are routinely missed for lack of any effective sanctions and no democratic polity has taken the responsibility of placing the issues officially and unambiguously before its electorate. (No polity anywhere has levelled with its public in that fashion, but democracy is supposed especially to promote civic honesty – and of course, many of the largest producers of greenhouse gas emissions are still democracies in one form or another.)

Nor is there any prospect that this failure will be redeemed over the couple of decades which we have left for effective action. Even under pressure of increasingly apparent climate destabilization and disasters like heatwaves, wildfires and flooding, we can expect democratic states to go on hesitating and temporizing, because their doing so stems from structural features of the liberal-democratic model – from the fixation of politicians on the short-term electoral cycle, the strong predominance of immediate (and overwhelmingly material) concerns in the minds of large majorities of their electorates and from the associated mass-cultural features which powerfully reinforce the natural tendency of these majorities to shirk complex issues requiring extended attention and sustained thinking.

But continuing failure is *not an option* – and that is not a form of words, a trick of emphasis, but the brutal and literal truth. Ensuring that we don't fail means fully empowering the properly informed and technically competent.

No such empowering will be initiated by democratic consent, because by the time the climate emergency is popularly recognized as that urgent, it will be far too late. It can therefore only happen if the informed and competent *take* power – their legitimation for doing so being, simply and sufficiently, the scale of the otherwise inevitable catastrophe.

This apparently obvious conclusion will be sturdily resisted, however, because underlying all the structural features which have ensured failure is the fundamental liberal-democratic given that we reach political decisions through finding a voting majority for them. Actions, policies and programmes, that is, are ultimately to be legitimated by head-counting. This approach, which would have seemed absurd in any age and culture before the eighteenth century, is one which people now find it enormously hard to see around – how else could legitimacy be conferred on collective decision-making? But this leads us to the second and deeper sense in which the 'undemocratic' objection is out of time. Head-counting can only legitimize decisions which essentially concern what heads have completely in common – where it makes *sense* to count one head as the equal of another. But it no longer makes such sense.

'All men are created equal', the classic Lockean statement of this approach, is not a universal truth but a declaration of intent belonging to a phase of political history which has now run its course. This is not simply because few now believe that men and women are created at all. The more modestly naturalistic claim that they are born equal is still not true, except when interpreted in the way which made sense during that phase. People are actually born with racial and genetic inheritances, physiques, mental capacities and potential abilities so hugely various that the claim of equality is nonsense unless it is taken to indicate that no one is any more or less his or her sovereign individual *self* than anyone else. This is indeed a truth (a conceptual one), and from it follows the practical consequence that each has the same interest in the things which go with being such a self – in individual well-being, freedom of action, scope for self-realization and self-development. These things then are what heads, or the selves which reside in heads, have in common. And when politics was all about how such interests were to interact and be adjudicated in the public sphere, as it was from the later eighteenth century until quite recently, and focused exclusively on the present because that is where such interests must be pursued and reconciled, it made a very clear kind of sense for every head to count for one, in Bentham's blunt idiom, and none for more than one.

But in the environmental age – that is, roughly since environmental issues emerged clearly to view in the final two decades of the last century – politics is no longer about reconciling the interests of presently-existing sovereign individuals. It has become, and for the foreseeable future must remain, the principal locus of our collective engagement with how the human species relates to its planetary habitat. This is now the master-consideration to which all questions of interests, individual, community or national, must be referred back. Policies and decisions must now be assessed with reference to their bearing on the health and stability of the biosphere as key to the planet's future habitability by the human (and a diversity of other) species – and most immediately, their bearing on the chances of surviving oncoming disasters in the near term and averting catastrophe further down the line. Insofar as the individual figures in this politics as such, it will be less as a sovereign self than as a manifestation of the wider and deeper life as it works within and through each of us – as what I have called a centre of life-responsibility. And in this regard, heads are not to be counted one by one, but *appealed to* in all their radically situated, embedded and embodied inequality. Empowering those heads (and hearts) which are capable of defending it is now life's only self-defence, and the legitimacy of pursuing that empowerment on behalf of life is unchallengeable.

It is true that head-counting democracy has brought significant benefits with it during its period of hegemony. It has led to improvement in material conditions for people at large, and (some) reduction in grosser material disparities. The establishment and defence of basic liberties for people in a wide range of polities has been an expression of the same spirit, as has a habit of respect for ordinary people just as human beings – the recognition, in R.H. Tawney's sardonic words, that 'even quite common men have souls'. It has also meant the supersession of government by aristocratic cabals (as devastatingly satirized for instance by Dickens in his portrayal of the Circumlocution Office) increasingly incompetent to manage the demands of post-Industrial Revolution societies. And it has been accompanied by the eventually universal extension of popular education, leading to raised standards of educated intelligence across the population (even despite the countervailing force of parallel developments in mass communications). All of these are important achievements which need preserving as far as possible, and there is no reason why many of them should not be preserved: material conditions are what must mainly change. Equally, the associated values (justice, solidarity, respect …) represent real human goods. That is why this situation is tragic, and must involve sacrifices which very many will find grievous. But this pain cannot outweigh the life-demands of survival, as objecting to a green vanguard's taking power to confront the emergency on the grounds that it would be undemocratic in effect allows it to do. Tragically – to repeat – we must now save the human habitability of the planet *at whatever cost*.

'Fascist'

The demand that biospheric health and stability become the master-consideration for political decision-making might seem to take us towards eco-fascism as defined by the environmental historian Michael Zimmerman: 'a totalitarian government which requires individuals to sacrifice their interests to the well-being of the "land", understood as the splendid net of life, or the organic whole of nature'. But to identify the aims of the revolution which is now needed in terms of such a definition would be seriously question-begging. Even leaving aside any doubts as to what human interests could possibly be served by climate catastrophe, such that they would be sacrificed in preventing it, that a government with such prevention as its overriding aim need be 'totalitarian' on anything like the model of the actually fascist European states of the earlier twentieth century simply does not follow.

Fascism as it arose in Italy and Germany at that time was a corporatist response to contemporary economic and political woes, including what was then perceived as the newly powerful threat of international communism. In the aftermath of the Great War this response became tied up with a resurgence of militaristic nationalism in both countries, and with an overtly racialist ideology in the case of Nazism. The only parallel offered by these historical developments to what will be needed in succession to the fossil-fuel state is the requirement of strong, indeed compulsive, central direction to set and maintain crucial parameters – carbon rationing, land conscription and a citizens' income, as already noted – which will certainly have a reshaping effect on the life of every citizen and the operation of almost every institution. But the introduction of measures with such far-reaching implications is a feature of any competent administration confronting an emergency, as is at least temporarily irresistible central direction. Churchill, for instance, reckoned that in 1940 the British War Cabinet had more effective power in Britain than Hitler had in Germany, and more recently the coronavirus pandemic has demonstrated worldwide to what lengths coercive state intervention must sometimes go to counter an existential threat. The power and reach of government in neither of these cases could sensibly be called a case of fascism. Yet the Churchill government imposed rationing and conscripted both individuals and industries to the war effort, while in the pandemic centralized income support and severe movement restrictions reorganized the lives of millions.

It is true that such forcefully transformative government, driven as I have been suggesting by a vanguard minority, would have to be authoritarian, at any rate in its earlier phases, and authoritarianism is very readily confused with fascism in this connection, and by the kind of objector to whom the terms propose themselves. But actually it is a very different matter, because it can involve the bypassing of normal modes of discussion and the compelling of acquiescence on justifiable authority. Nor does justification have to arise, as of course it did in those two examples, ultimately from accountability to representative assemblies at least notionally democratic. While a politics of interests focused on access to resources and 'the pursuit of happiness' remains the default mode, claims to authority derived from any other than majoritarian and egalitarian arrangements are naturally regarded as suspect – they are taken to mask the pursuit of control by some group (the robber capitalists behind Putin, the self-recruiting bureaucracy of Chinese communism, the international financiers enabling Trump) whose particular interests are inimical to those of the majority; and where interests are indeed the main issue, the wider should evidently take precedence over the more particular, so that 'authoritarian' becomes automatically a derogatory label. But as we have seen, the politics of interests no longer serves or defines the environmental age; and when the future habitability of the planet is at stake, democratic process is palpably failing and the green vanguard bases its actions on hard facts which the majority are disposed to ignore to the existential peril of all, its authority involves no bogus claim but asserts and seeks to implement the inherently unchallengeable authority of *truth over falsehood*.

Notwithstanding that, it is also undeniably true that in the present fragile state of liberal-democratic politics, deliberate subversion of the fossil-fuel state even in the name of ecological truth might open a way, through some ensuing melée, for bogus authoritarianism of a reprehensible kind (perhaps genuinely fascist or quasi-fascist) – or maybe for the undemocratic rule of some interest group (the military, for instance) which saves the human habitability of the planet by the way, as it were, in order to consolidate its own grip on power. Here, though, we are brought face to face with a straightforward balancing of risks: that of either of these eventualities, against that of a dying planet and possibly human annihilation if decisive action is neglected. It seems to me, though it may not to others, wholly clear which of these risks we should prefer to run. What is quite unhelpful is pre-emptively to confuse readiness to run them itself with fascism.

Moreover, there is every reason why a state whose ecological parameters were set centrally and authoritatively by revolutionary green-vanguard action should still preserve a wide range of effective democratic procedures at local level where the impact of the necessary changes on actual patterns of economic and social life will have to be negotiated. Here the need for a fair reconciliation of individual interests and the corresponding claim to a working form of equality will continue to be prominent, and decision-making structures will have to operate as far as possible through popular consent and endorsement – it is only the fundamental reset of the state which cannot now wait for this kind of validation. The sinews of a more vital and participatory local democracy which will thus have to be strengthened and nourished have no parallel at all in historical fascism with its inherent centripetal tendencies and relentless drive to top-down coordination.

As to the most destructive and dangerous elements of the classic fascist package, these are simply irrelevant to the present situation. Although, as things are, revolution can only remove the fossil-fuel state country by country, it must do so in the service of a radically internationalist vision, since climate chaos is no respecter of national boundaries and will render any purely nationalist response impotent. (It is true that the 'blood and soil' strand of racist nationalism under Hitler was articulated partially in ecological terms, but this line of thought goes back far beyond Nazism within broader German culture, to the technical forestry concept of *Nachhaltigkeit* coined in the early eighteenth century which has latterly mutated into the necessarily global idea of sustainability.) Again, climate breakdown will generate many armed conflicts over essential resources in the coming decades, but it is not open to being *addressed* by military adventurism of the kind historically inherent in the fascist dynamic. Nor is there any room for drumming up fervour by targeting for blame and abuse, still less by arming against, some 'enemy within', be it communists, immigrants or any other supposedly alien group. The enemy in the climate crisis is always and only ever ourselves, in the form of general lock-in to a destructive mode of life originating with Western capitalism but now globalized. The crisis being for that reason alone so wholly unprecedented, inept labels like 'fascism' merely serve in this context to darken counsel. They should be dismissed wherever they are encountered.

'Elitism'

Charges of elitism, on the other hand, should not be dismissed, but embraced – because only a little thought is required to make it plain that this attribution is not really a charge. Or rather, it can only constitute a charge when the word is used in a certain sense, but the sense in which it applies to the vanguard role here being canvassed is a different one, so that objecting to that role on that basis exemplifies the fallacy of *ignoratio elenchi* or arguing beside the point.

To be clear: whether the existence of an elite group in any particular field is a bad or a good thing, and so whether 'elitism' as support for that existence is a matter for blame or praise, depends entirely on the relation between the nature of the field and the criteria by which the group in question is *chosen* (French *élite*, whence the term). Elite groups and arrangements are to be reprehended whenever these criteria are essentially irrelevant to the field – as for instance with 'elite educational establishments', entry to which is either determined or decisively influenced by wealth and social status rather than exclusively by capacity to profit from a high-quality curriculum; or as with governing elites filled predominantly by the products of such establishments. By contrast, the emergence of an elite group of performers in some sport, through talent-spotting followed by appropriate training regimes, and the subsequent selection of members of that elite for teams entering top competitions, is not to be reprehended at all but applauded (by those interested) as maintaining and improving levels of performance and aspiration in the given activity.

In the political arena, there is obviously a connection between elitism in the reprehensible sense and the growth of explicit or tacit oligarchies – ruling groups which batten on power while recruiting themselves principally from people like themselves, so that the key criterion becomes ability to conform within a self-reinforcing and self-perpetuating structure rather than capacity or proper authority for the tasks of government. But the green revolutionary vanguard has nothing of this character. As described when we were considering its necessary qualities, it is chosen specifically on the basis of its capacities, and chosen to take on a responsibility which must be discharged by those qualified if we are to avoid climate and ecological catastrophe.

As to chosen *by whom*, the whole account which I have offered up to this point should make it apparent that they are really 'chosen' by the life-impulse itself. To make what is fundamentally the same claim from a different angle, they are 'self-appointed' in essentially the way that the creative artist is self-appointed – the way which insists on itself as clearly valid as soon as we ask: who *else* could conceivably appoint him? And this selection is according to living criteria which could not be more urgently relevant. Hence in its quality of an elite the vanguard deserves only respect, and acknowledgement of its claim to authority on that basis is not an objection, but a recognition of what might still, if we are both brave and lucky, save us.

I cannot, indeed, forbear concluding these considerations about democracy, authority and elitism on the very opposite of a defensive note. We in the Western states where the crisis of our time will be decided are now confronted, uniquely in history, by a commodity-addicted majority populace whose understanding has been pawned into the keeping of a corrupt and contemptible gutter press, whose thinking has been fragmented by Twitter and the associated solicitations of instant opinion, and whose general living has been exiled from the natural world by urbanization and robbed of human depth by the destruction of all but the helpless vestiges of a humane culture. The case for the last conjunct here has not been made as such in this book, but only glanced at in passing, and its proper development must be reserved for another place; but the travestying of life-purpose and meaning by commodity-addiction has been sufficiently invoked, and that condition could not have become so widely prevalent except in the absence of any strong humane culture to preclude the gap in the soul for which commodities offer a substitute filling. This is the counterpart of existential immiseration felt by the minority alert enough to their life-depth to recoil from complicity in biocide: it is the *unfelt* condition of a majority cheated of their real lives but numbed to their loss by addiction. Arguably – and it is an argument which all important political thinkers from Plato onwards would have recognized – such unprecedented spiritual impoverishment means that those afflicted have forfeited any overriding claim to govern themselves. Even if that be thought too strong, they have certainly forfeited any claim to arbitrate the human future, including whether or not there will be one – which comes, where climate crisis is the overwhelming political issue, to the same thing. That bleak, tragic but surely unignorable situation must be the charter as well as the challenge for the newly life-responsible vanguard as it refashions itself for its novel duties. Members of that vanguard are now representative of the whole of humanity in a way which altogether transcends the statistical. They *are*, in fact, for this purpose humanity in its wholeness, rejecting two-dimensional commodity-addiction out of the depth of life and on behalf of life. That is their justification for action, and (if beyond morality we are still to use rights-language) the ground of their right to exercise whatever power they can lay hold on for trying to avert catastrophe. They act on the authority, we might say, not only of truth but also of life. They neither can nor need await majority endorsement in any form, before taking all measures open to them to urge, assist, marshal and as necessary compel that majority to save itself.

And majority salvation in the longer run is, of course, fundamental. It is implicit in such a charter – and another vital difference from any authoritarianism in the bad sense – that the vanguard should pursue on its living authority transformation which both permits and encourages recovery of human wholeness for all. The goal must be to renew a situation where, in Rainborough's ringing words from the English Civil War, 'the poorest he … hath a life to live as the greatest he', and none is betrayed by a perverted civilization into a destructive travesty of life. By the same token, vanguard-driven transformation is only likely to take hold to the extent that its initial imposition releases some very significant measure of early popular endorsement – which one might imagine as a gathering wave of widespread, horrified revulsion against commodity-addiction as the structural conditions pressurizing people into it begin to be removed. Our hope against hope lies in the fact that this sort of response cannot be ruled out, although it really does test one's understanding of the term *unprecedented* – there have been revolutionary surges demanding political rights, material benefits, religious freedoms … but never yet an existential revolution. If that wave of revulsion, ever-strengthening, took the form of recognizing the twofold authority of truth and life in the saving vanguard, and an accompanying readiness to accept their direction for the purposes of confronting the emergency, we might still be in with a chance, and humanity might still have a future.

A revolutionary programme

If the green vanguard is led inexorably by its qualities of intelligence and character to take up a revolutionary stance, is potentially well positioned in a complex techno-bureaucratic society to pursue power and is also fully entitled by the nature of the emergency to take power if it can, the question of course remains whether there is a feasible programme to be envisaged for its doing so. Does the vanguard have a route to power which might be plausible, or at any rate not wholly implausible?

The latter qualification is not merely a matter of hedging one's bets. As this book's argument should have made plain, the domain of the *not wholly implausible* is where we are now at best for all of this: it is part of our plight that we must invest whatever hope we can summon against very heavy odds, as our only chance of changing the odds. The odds against any revolution are always heavy, in the nature of states, institutions and political habits (and also perhaps of revolutionaries). Against this vital revolution, they are heavier than humanity has ever faced. That recognition must not daunt us, but it should guide us – towards, crucially, a possible programme of multiple strands, operating at a variety of levels and capable of being pursued with vigorous opportunism and great tactical flexibility.

I think such a programme and route to power can be identified, and in these concluding sections of the book I shall sketch what I take them to be. That this is no more than the barest sketch cannot be overemphasized. To develop any of its lines of suggestion in anything like operational detail would require at least another book – preceded, ideally, by a substantial period of diverse and energetic experimentation. All I try to do here is indicate, in an extremely outline and provisional way, the directions in which the general reorientation of our political thinking promoted by tragic understanding might take us. I do this, as I have stressed, as a basis for the debate which I hope to provoke and in the firm belief that we have an obligation at least to explore the possibilities. My sketch is written with the British context specifically in mind, since it is that with which I am familiar, but the suggestions are meant to apply anywhere that a vanguard movement of the kind which I have described could come to consciousness of itself as confronting a fossil-fuel state.

This is the point at which to say something more about the reference of that shorthand term *fossil-fuel state*. To the extent that they are caught up in a worldwide capitalist economy heavily reliant on energy from fossilized carbon, very many states – from the medieval tyranny of Saudi Arabia, through oligarchical kleptocracies like Russia all the way to the Western democracies – could be so labelled. But I want to reserve the description, and use the term in what follows, for states which are *structurally* tied in to fossil-fuel dependency, rather than just swept along with globalization. These are principally the liberal-democratic polities in which protecting and facilitating individual freedom to pursue the good life is taken to be the state's fundamental point and purpose, and where in consequence the carbon-based, energy-intensive material consumption now widely misperceived to constitute the good life has become inseparable from their very political rationale. The export of these destructive expectancies to other kinds of state (India and China being prime examples) is now of course a major driver of global emissions: but it is these Northern and Western liberal democracies – and first of all, historically, Britain and the US – where the disease originated and is now endemic. By the same token, they must be the principal targets for the revolutionary impulse and dynamic which we have been tracking. For a state is ripe for revolution when it is its structural arrangements, those essential to its constitution and self-understanding and too deeply entrenched for adaptive change, which produce the immiseration giving rise to transformational energies.

The programme which I shall outline is threefold. Its first strand, as I have already indicated, involves delegitimizing, destabilizing and finally rendering inoperable the fossil-fuel state. This needs to be undertaken in each respective jurisdiction – for British readers, here in Britain in the first place – although, of course, the process must be rapidly exported worldwide from wherever it achieves success. The second strand is to put in place a shadow eco-state to take over when the fossil-fuel state disintegrates. In parallel with these two proactive categories of endeavour, the third strand might be described as reactive, but it is at least as important: it is that of insistently emphasizing and publicizing *as* climate-driven the disasters (wildfires, floods, pandemics …) which will increasingly occur over the next two decades, in order to educate the public at large into at least acquiescence when revolutionary political and economic change is forced upon them.

I will consider each of these three strands in turn. The demands of summary and compression will often no doubt give this consideration an appearance of dogmatism, but I stress yet again that nothing of the sort is intended: I am attempting no more than to sketch a possible framework and stimulate debate.

Destabilizing the fossil-fuel state

Under this head, mass civil disobedience of the kind which XR has been developing will of course continue to be very important – not least because such actions, as well as bringing capital cities and governing districts to a series of temporary halts, play a vital role in both educating and engaging the wider public, demonstration by demonstration. Not everyone whom these demonstrations get out onto the streets need be part of the initiating vanguard minority driven by full consciousness of its existential immiseration, but many of them will be – more especially among those willing to risk arrest – and the power of this self-sacrificing determination to recruit increasing numbers of people to at least a proto-revolutionary understanding of the climate emergency will continue to matter vitally.

If the scepticism expressed in Chapter 7 about XR's own announced route to power (increasing the size and frequency of its demonstrations in order to push government to establish Citizens' Assemblies with delegated legislative authority) is warranted, however, a much more focused strategy will also be needed. In particular, the concept of the *climate strike* will need to be taken to a new level.

When we asked where vanguard members were likely to be found, the answer was: in many walks of life, but perhaps especially in the professions, in administration, in communications and in the management of the more 'conceptual' industries. While not having the kind of power which goes with being a governing oligarchy or an economic class, in other words, they will most of them have their hands on or near the levers of both efficacy and legitimacy for the fossil-fuel state: we noted that without their cooperation and their tacit endorsement, that state could not go on functioning for any very long time. The point is then to recognize and use that potential leverage in a properly organized way, so as to bring targeted pressure to bear wherever and whenever the structure of the fossil-fuel state starts to show cracks. This will require the formation of 'vanguard cells' across the broadest possible range of professional and occupational groupings, and the coordinated and increasingly frequent withdrawal of cooperation and endorsement by members of these cells over a significant period of time. In this process, the new possibilities offered by social media for group formation, liaison and both dispersed and centralized communication must obviously play a key role, to the extent that the platforms used can be secured against surveillance by state forces.

Thirdly, within this category (and here with the UK political context especially in mind), there will certainly be a continuing role for a green political party dedicated to contesting elections. Electing people to the local councils which will have to cooperate with the localized eco-state as it emerges (for which, see later) will remain important. Nationally, however, this activity would need to be concentrated in a very much more disciplined way than has yet been the case on a small number of Parliamentary seats which may become winnable as the public mood changes in response to accumulating disasters. Crucially, this must also involve the getting elected in those few seats, not just of anyone local who generously proffers themselves, as has been the tradition of the Greens in the UK, but of potentially powerful Parliamentary performers who could help to form an effective blocking minority in the event of a hung Parliament, and who would also be capable of taking key ministerial roles (crucially including police and internal security, as well as 'environment') in a government of national unity, should one be formed as part of addressing the emergency. For such a stage in the disintegrative transformation of the fossil-fuel state cannot be ruled out as the pressure mounts, and the capacity to exploit any situation of this kind through Parliamentary as well as extra-Parliamentary means must be cultivated as part of any suitably flexible revolutionary programme.

And what about more drastic extra-Parliamentary means than those endorsed by XR? Speaking personally, I am strongly in favour of non-violence – like, I imagine, most sane people, I had far rather be on the receiving end of it than of the alternative. It is also evident that the committed practice of non-violent civil resistance can be a powerful force for bringing into the movement people who might not otherwise join up. All that is very understandable. But a strategy of targeting, for instance, key fossil-fuel state infrastructure such as motorway bridges and airport runways, backed by the necessary minimum of force if challenged, would also be a very understandable reaction to the desperate plight which we are in. It is one which has already been canvassed in a US context under the rubric of 'deep green resistance' – and one which a state using the law to promote biocide has squandered any right whatsoever to condemn. The difficult issue then becomes one of coordination. You cannot have a single movement with an enforcing as well as a non-violent arm – the concept of non-violence just seems to rule this out. By the same token, Parliamentary activity is going to be obliged to disavow this kind of challenge to established power. But there are historical precedents for revolutionary movements which have inspired in parallel, as it were, and proceeded through different modes of action by different groups of people, and we could perhaps learn things from those histories. Here it is most apparently tragic that after a century and a half of largely peaceful politics (at least in Britain), we have now to think again in these terms. But it is also clearly irresponsible to shirk doing so through a settled refusal to recognize tragedy when we see it, given that so much is at stake.

Building the shadow eco-state

If the fossil-fuel state has its political rationale tied to a high-carbon economic and social structure, the eco-state is its antithesis: structurally committed to zero carbon and sustainable impacts on the biosphere, its political rationale – one unseen hitherto in history, but henceforth paradigmatic if humanity is to survive – will be ecological balance to the extent possible within its own jurisdiction. That we know already quite well enough what the main characteristics of such a state must be, is the real standout achievement from forty years of green politics. Its most striking differences from its fossil-fuel predecessor will flow from the drastic reductions which it will have to enable, or enforce, both in discretionary personal mobility over long distances and in the product-miles embedded in ordinary everyday consumption.

Absolutely crucial to these reductions will be the development of genuinely localized economic resilience. And here, not necessarily for reasons to do with the climate emergency, civil society has itself begun to anticipate the necessary revolution. There are already in existence a considerable number of the components from which the local structure of the eco-state will need to be built. These include Transition Town organizations and other local sustainability projects, local food and other production and market arrangements, local currency set-ups, renewable energy schemes such as community wind turbines, community businesses supplying a variety of services, shared-car schemes to replace or supplement inadequate public transport arrangements and many forms of neighbourhood self-help, sometimes organized by recognized bodies like community centres and Town or Parish councils and sometimes arising more or less informally.

These various civil-society activities represent strands of local self-reliance, presently disparate, which will have to be knitted together opportunistically to form the sinews of the locally organized economies, supply arrangements and governance structures on which we will come increasingly to depend. They are the building blocks for subsidiarity created from the bottom up as the globalizing project fails. Some of them, for instance Transition Towns, are already largely aware of themselves in essentially that role (or the not dissimilar one of preparing for 'energy descent' following 'peak oil'). Others may have arisen to address specific needs in response to the retraction of public functions by the liberal-capitalist state, originating with Thatcherism and rapidly accelerated under post-2010 austerity; still others may simply be traditional community self-help initiatives with as yet no thought of any overarching political rationale for their work. The role of the green vanguard must now be to draw all these initiatives together area by area into the outline form of the shadow eco-state at ground level, ready to ensure that resources and services are in place locally as the globalized economy unravels and the fossil-fuel state withdraws to the essential national-level functions of defence, police, upholding the legal and fiscal systems and maintaining key infrastructure.

So far this kind of work goes with the grain of much already-existing green-movement activity. But there is a vital further dimension to assembling the shadow state which has not yet emerged and will need to be newly developed. This involves the formation and keeping actively in play of a shadow central government, continually tracking, harrying and challenging elected government (and official opposition) while the fossil-fuel state lasts, and constantly foregrounding the green alternative. And that means much more than maintaining in being an electorally focused Green Party which issues periodic press releases drawing attention to various policies elaborated in its latest 92-page manifesto. Indeed, it means boldly abandoning fantasy manifestos altogether and aiming for a Leninist cogency and simplicity. 'Peace and bread, the land to the peasants, the factories to the workers' were the slogans which made the Bolshevik revolution possible, and the Green equivalents – 'Carbon rationing, land conscription for local food security, a living-wage citizen's income …' – are what should be ceaselessly hammered home to accompany each exposure of fossil-fuel state inadequacies as the climate emergency gathers pace.

As a vital part of the shadow government operation, a network of relevant expertise based on the vanguard cells suggested earlier should be established and tasked with studying and preparing the measures needed to turn those slogans into a functioning economic and political framework. But this 'shadow civil service', as we might well think of it, while preparing to support a transitional or emergency administration, should certainly not be trying to second-guess how these key revolutionary changes will play out transformatively through economy and society. No one could predict this in any kind of detail, and therefore no one can credibly offer to *manage* this process. Here again the tragic understanding emerges as a keynote of appropriate realism. The dramatic shifts in patterns of production, consumption and mobility needed to reach net zero emissions in time to stave off catastrophe, while their effects are unpredictable in detail, will certainly bring with them huge corresponding changes in broad patterns of employment, income and all related economic activities – changes which will not accomplish themselves without major disruption and loss (and not just of comfortable habits). The building up of local resilience combined with the introduction of a citizen's income – long overdue, and now with the COVID-19 crisis demonstrably something for which the state can find resources – should help ensure that few people actually starve during this shake-out. But many will suffer, some of them grievously. Once we have decisively left behind the utopian aspiration to save ourselves painlessly, however, this can be recognized as the inevitable concomitant of doing what the emergency calls for.

Public education through disaster

Climate-driven disasters will occur with increasing frequency over the next couple of decades, whatever we do – they are inevitable given the warming trajectory to which humanity has already committed the planet with its greenhouse gas emissions to date. We should not underestimate the effect which this series of disasters, properly interpreted, might have on shifting the public mind towards acceptance of transformative change. But neither should we overestimate what that effect might be without interpretation, nor misconstrue what proper interpretation will involve.

The COVID-19 pandemic, still raging as I write, offers a useful test-bed for the issues here. This phenomenon is neither climate-driven nor directly climate-related, but it has an anthropogenic origin and aetiology comparable to those which the disaster-sequences of climate emergency are already exhibiting and will display even more floridly as time goes on. According to the World Health Organization this particular disease appears to have originated from a market in China where wild animals, including marmots, birds, rabbits, bats and snakes, were being traded illegally – in a clear instance of consumerist human over-reaching and disrespect for nature. Coronaviruses jump from animals to humans, and it is thought that the first people infected with the disease – stallholders from the market – contracted it from contact with some of these animals, possibly the bats. It then extended its reach worldwide as a result of globalized hypermobility (spreading at the speed of jet planes), with its infectious power intensified by crowded urban living and its destructiveness by the lack of resilience in centralized health and welfare systems. In the process, it has hugely disturbed economies and living arrangements in a way which hitherto could only have been predicted of war or climate crisis. So what can we learn from this experience to help us deal with the latter?

Even with the pandemic still in full career, some voices are being raised to claim a potential COVID-19 bonus for climate campaigners – an unexpected last chance for conventional (that is, non-revolutionary) politics to rise adequately to the climate emergency. The broad public acceptability of stringent restrictions on mobility (with significant knock-on effects on patterns of consumption) as a government-imposed and science-driven public health response to the COVID-19 emergency may carry over, it is being suggested, once that emergency is defeated or at least contained, into a readiness to accept comparable restrictions necessary to hit zero emissions targets. Recognizing that driving less, shopping locally, working from home and developing or intensifying community solidarity have worked in helping to stem COVID-19, people may then be readier to bring these and similar lifestyle changes to bear in the service of addressing the climate emergency.

These specific anticipations are speculative at the time of writing. Human nature may surprise us – but by the time these words are in print, we may well have found out just how powerfully many people had missed driving and Metrocentres, how much more exacting they had found working at home (especially with family around) compared with escaping to the office and how actively they had disliked depending on the sporadic benevolence of their neighbours. If so, that will surely be because to have expected otherwise was to have ignored the *radical* difference between the COVID-19 and climate emergencies. This is that COVID-19 kills indiscriminately and in the full glare of media attention – so that everyone is liable to be scared by the immediate possibility that they or their loved ones will die of it, and in some cases too, more altruistically, by the prospect of widespread illness and death accompanied by the collapse of key public services on which all rely. The upshot of this general fear has been widespread acceptance, at least temporarily, of lockdowns and related public health measures. But the climate emergency and its implications only scare *a minority* of people – those, as we noted earlier, who couple honesty and courage with the intelligence to understand and heed the science, the imagination to bring the predicted impacts of global overheating vividly before their minds and the reflectiveness to realize that trying to shrug these prospects off and concentrate on one's present life robs the latter of any meaning or point. So, in the immediate future at any rate, there will only be 'climate-driven disasters', testifying to the existence of a frightening climate emergency, *for* that minority: majority readiness to confront them as such is not to be expected, and nor therefore is any general readiness to accept the restrictions and rearrangements going with an actually recognized emergency.

Now of course, the boundary between this minority and the rest is not fixed. The COVID-19 experience may be found to have helped strengthen understanding in those where it had been only potential – in particular, by highlighting the benefits of attention to the consensus of relevant experts. It may have enriched imagination through encounter with the hitherto unimaginable; and it may have prompted the kind of reflectiveness about life-meaning in the face of existential threat, openness to which just goes with being human. The same may well turn out to be true of each subsequent anthropogenic disaster, both the more and the less clearly climate-related, so that the constituency of acceptability for the necessary socio-economic transformation is successively broadened as time goes on. Similarly, collective response to these disasters may cumulatively promote the rediscovery of various forms of community solidarity and the revitalizing, strengthening or in some cases creation of infrastructures of local resilience. This renewed social capital for the eco-state may well be carried forward and augmented as further dangerous events unroll.

The key point, however, is that what both these forms of 'disaster bonus' depend on is deliberate strategic activity on the part of the vanguard minority to whom the climate emergency is real. Any kind of social learning from disaster as we go forward is going to involve acceptance by that conscious minority that they must now be proactive without waiting on widespread popular consent. They must insist relentlessly on the climate connections of each successive flood, storm, wildfire or similar event; they must constantly reinforce the discourse of emergency and hammer away at the slogans of transformative response. (They can certainly emphasize that implementing these slogans would involve arrangements much less restrictive to the individual, and even perhaps in sum less economically disruptive, than the COVID-19 lockdowns have been.) But they must also make immediate moves and preparations which, before there is widespread public consent, will be understood as revolutionary – trusting that sufficient consent or at least acquiescence will follow as even the residual legions of the thoughtless are forced to acknowledge at last what is happening. Waiting until climate emergency is as widely recognized as was the COVID-19 danger – until, say, it is headline news in the *Daily Mail* – before starting the revolution would be to have left things helplessly too late. But recognition just *may* come in time for acceptance of a revolution already substantively in train.

A serious Green politics?

How does the existing political arm of the Green movement – the vanguard already *en marche*, as it were – relate to all of this? Here, I focus on the Green Party in Britain, though much of what follows will also be relevant to its sister parties in other parts of the world.

The Green Party has at the time of writing a membership of around 50,000 people across the UK, the great majority of them in England and Wales (notionally separate and much smaller party structures operate in Scotland and Northern Ireland). Members are organized in over 200 branches or locality Green Parties, which function with a large measure of autonomy from a nevertheless rather Byzantine hierarchy of regional and inter-regional committees, with a Party Executive comprising co-leaders and coordinators for various tasks sitting somewhat uncomfortably above the inter-regional tier. The main business of the membership of this considerable if unwieldy-looking political force, apart from general publicity and campaigning on green issues locally, consists firstly in selecting and attempting to get elected candidates for local, national and (until recently) international representative bodies; and secondly, in an ongoing debating process out of which the Party's bank of policies is continually being created, revised and refined.

Electorally, these activities have been only modestly successful, resulting at the time of writing (September 2020) in one member of the UK Parliament (three if one counts a couple of peers), six of the Scottish Parliament, two of the London Assembly and two of the Legislative Assembly in Northern Ireland – plus something under 400 seats on principal authority councils nationwide. (Until Brexit there were also seven Members of the European Parliament.) The articulacy and effectiveness of these representatives is often disproportionate to their small numbers, but their voting strength of course isn't. Policy formation, however, has proved conspicuously more successful, at any rate if success is to be measured in terms of happy activity and pages filled up. The foreword to the Party's Political Programme boasts that 'Our policy is set by our members – meaning that our leaders have no more voting power than our newest recruits', and a regular succession of two conferences annually (which any member can indeed attend) has so far produced a book-length portfolio with chapters of detailed prescription ranging from Animal Rights to Workers' Rights via most of the alphabet in between. From this wealth of material is extracted periodically the manifesto for each election in which the Party fields candidates, and also the standing

Political Programme just mentioned. Manifestos and Programme include proposals for introducing binding sustainability and environmental targets, reducing energy consumption, encouraging renewables, restoring the public transport network, supporting the development of green jobs and introducing a citizen's income. (Most of these were encapsulated pragmatically for the 2019 General Election manifesto in summaries of ten Bills to be enacted in the first two years of the new Parliament.) They also include a range of aspirations for mandatory realignment of pay differentials across the economy, gender and diversity quotas for every type of organization, free higher education for all, acceptable forms of relationship instruction in schools and regulations concerning pet ownership.

And what's wrong with that profile? Regretfully, in the light of all the foregoing, the answer has to be: roughly, everything. Regretfully, because I am myself a long-standing member of this organization and can testify to the idealism, the unsparing commitment and general good-heartedness which goes into all its activity. The Greens have come a very long way since we were the Ecology Party with a few hundred members and a national operation run out of a friendly living room; trying for forty years to make this approach through electoral politics work has been a noble enterprise. But – at least in the form just described – it is one whose time is now past. The Party garnered some 850,000 votes in 2019, its highest ever total in a general election – and that still amounted to only 3 per cent of the poll. With at best two decades to achieve net zero emissions if climate catastrophe is to be averted, the coming to power of the green programme through any purely electoral process is now evidently much too high a mountain to climb, and this would remain true were a sensible system of proportional representation to be introduced for all UK elections tomorrow, and even if coalition-building around a Green New Deal were a more hopeful prospect than I have argued. The principally electoral route for green politics – the opportunity for decisive influence through elections fought as anything more than one small component of an overall revolutionary programme – must now be recognized as locked and barred.

And if that is the case, then indeed everything which the current Party does really belongs with all that earnest frolicking through the policy meadows, as a kind of displacement activity: a way of avoiding confrontation with the reality of our plight. The Political Programme document speaks of the Green Party as 'an insurgent force for good' engaged in making 'the desirable feasible'. But what it is actually doing, in persisting along its present track, is making the well-intentioned comfortable – and correspondingly, the undesirable unchallengeable.

Or rather, that will be its cruel epitaph unless it now very rapidly adapts itself to claim its place at the heart of the kind of programme which I have been sketching in this chapter. For it is clear that such a programme needs above all a coordinating political protagonist, a strategic focus, nerve-centre and general staff. In this absolutely vital function the generously welcoming inclusiveness and decentralized looseness hitherto cultivated by the Greens will simply not serve. Rather, there must emerge from this creatively spontaneous matrix the explicitly revolutionary green-political party for which the situation now so vehemently calls.

If greens are seeking a serious political model adequate to the times, in other words, they should be picturing themselves no longer as the thinking person's Liberal Democrats, but as something much more like the Bolsheviks. I do not choose that comparison merely for its shock value: there is an important point of what might be called political epistemology behind it. What enabled the Bolsheviks to capture the Russian state in 1917 – apart, of course, from their unrivalled organization and firmness of purpose – was, in very large measure, the driving force of Lenin's absolute conviction that Marxism was a genuine science, according to which the inevitability of proletarian power was just *how things were* in the world, a historically determined necessity which the October Revolution offered the swiftest way to realize. That coup turned quickly to terror and tyranny because, of course, Marxism was actually pseudo-science, and the supposedly inevitable dictatorship of the proletariat rapidly became the dictatorship of power-hungry personalities. But the Green claim to reshape society and create a successor ecological state is based on the near-unanimity of genuine science to the effect that the conditions for inevitable climate disaster are now firmly in place, and also on offering the only plan for reorganizing society and economy which even begins to face up to that reality. Green revolutionaries, that is, would be both worse and better placed than were the Bolsheviks. We have a heavier responsibility, to transform the world almost out of recognition in very short order indeed – but we really do know how things are, if only we dare to act on that knowledge.

Still, the emergence, out of what we have now, of a green revolutionary party with the leadership strengths and the organization, focus and discipline which would be needed, will require a cultural metamorphosis in the Green-political movement at least as far-reaching and dramatic as the lifestyle changes for which we must go on hoping from the population at large, and also (as befits a vanguard) happening very much sooner. Is green politics, are Green politicians, up to that challenge? Can they become crucial agents of the tragically-informed transformative hope which this book has been about demonstrating to be still, even yet, realistic?

Much – actually, *everything* – hangs on the answer.

***

I expect this chapter to have been met with growing incredulity or resistance as it has unfolded towards that conclusion. Had it been a lecture, there might now prevail an uneasy silence, broken here and there perhaps by angry muttering.

For those to whom the climate emergency is not yet real (should any have stayed the course), its suggestions will have appeared as fantasy, uncalled-for and only not dangerous because so wildly implausible. But for most of the initially sympathetic, among whom some may have found the earlier chapters persuasive, I will be felt to have concluded with a direct and very disconcerting challenge to everything that they have been used to believing about the nature and prospects of Green politics – indeed, of politics and social action generally. So what chance can I suppose the book as a whole to have of helping to change the way we do things over these crucial coming decades?

The book's own line of argument offers the only possible reply. Empirically, it might be realistic to suppose, no chance – but while life remains, there is also hope against hope. And it attaches in this case to the possibility that even just one or two present or potential leaders of thought and action could concede my premises, find the logic compelling and then be bold enough to accept that the conclusion, however shocking and disruptive, really follows.

Vanguards themselves have to be led, revolutions must start in creative individual minds and life has deep resources at which we cannot guess. That, at any rate, is the belief – the hope – in which I have written.

Notes

For full bibliographical details see References, listed at the end of these Notes.

Introduction

p.1 ***Epigraph***: from the last stanza of 'Prometheus Unbound' in Shelley (1967), p.268.

p.2 ***the 70 per cent increase in global demand for coal***: see Helm (2012), p.32.

p.2 ***China … may be planning further major expansion***: see *The Guardian*, 20 November 2019, available at: https://www.theguardian.com/world/2019/nov/20/china-appetite-for-coal-power-stations-returns-despite-climate-pledge-capacity

p.2 ***the twenty years which we maybe have left***: at a generous estimate – the IPCC 2018 Report (available at: https://www.ipcc.ch/sr15/download/) suggested twelve years at that point.

pp.2–3 ***Extinction Rebellion … the school strike movement***: available respectively at https://rebellion.earth/the-truth/ and https://ukscn.org/ys4c

p.3 ***Greta Thunberg***: see Thunberg (2019), p.24.

p.3 ***My own work in this field***: see in particular Foster (ed) (1997), Foster (2008) and (2015).

p.4 ***all the consequences of a 4°C rise***: see, for details, the IPCC report cited previously.

p.4 ***to hope for X***: this is the accepted standard philosophical account or 'orthodox definition' of the concept. See for example Martin (2013), p.11.

p.6 ***very well worked over in that sort of book***: I am thinking here especially of Jackson (2009); Klein (2014); Monbiot (2017); Porritt (2020).

p.7 ***David Attenborough***: see Attenborough (2020) – the quotations are from pp.220 and 217–18 respectively.

p.8 ***'step up or step aside'***: quoted in Porritt (2020), p.63.

p.12 ***conceptual work ... as a specific intervention***: this role for, or particular way of understanding, political philosophy is discussed under the head of 'conceptual innovation' by Raymond Geuss (2008), pp.42–50.

p.12 ***Thomas Nagel***: see Nagel (1989).

Chapter 1

p.14 ***odds ... against your being struck by lightning***: see, for example, https://discovertheodds.com/what-are-the-odds-of-being-struck-by-lightning/

p.15 ***George Orwell's classic insight***: see the article 'In Front of Your Nose' reprinted in Orwell (1970b), Vol 4, p.154.

p.17 ***'the vicious syllogism'***: see Foster (2015), p.2.

p.18 ***The current scientific consensus***: for summary and references see the website of the Carbon Tracker Initiative at: https://carbontracker.org/carbon-budgets-where-are-we-now/

p.19 ***the Carbon Countdown website***: available at https://www.theguardian.com/environment/datablog/2017/jan/19/carbon-countdown-clock-how-much-of-the-worlds-carbon-budget-have-we-spent

p.20 ***The carbon cliff-edge***: image from Berners-Lee and Clark (2013), reprinted with permission from Profile Books and Greystone Books.

p.20 ***Professor Kevin Anderson ... has argued***: in Anderson (2012), available at: https://content.csbs.utah.edu/~mli/Economics%207004/wnv3_andersson_144.pdf

p.21 ***no credible scenario under which all this indirect energy demand is shifted to renewables***: see again Helm (2012).

p.21 ***As Anderson soberly puts it***: Anderson (2012), p.26.

p.22 ***a wide diversity of authors***: alphabetically: Jem Bendell, 'Deep Adaptation', available at: https://jembendell.com/category/deep-adaptation/; Hamilton (2010); Jamieson (2014); Mulgan (2011); Read and Alexander (2019); Wallace-Wells (2019).

p.22 ***the 'Anthropocene' trope***: see, among much else, Hamilton (2017).

p.22 ***as Brian Heatley and Rupert Read point out***: in 'Facing Up to Climate Reality: Introduction to the Project', available at: https://www.greenhousethinktank.org/uploads/4/8/3/2/48324387/intro_to_fucr_project_2017_spring_conference_edition.pdf. See also in this connection Anderson on carbon capture and storage: http://kevinanderson.info/blog/the-hidden-agenda-how-veiled-techno-utopias-shore-up-the-paris-agreement/

p.22 ***described by those in a position to know***: for example by Anderson (2012), p.29.

p.23 ***a 'perfect storm' of food, water and energy shortages***: see *The Guardian*, 18 March 2009, available at: https://www.theguardian.com/science/2009/mar/18/perfect-storm-john-beddington-energy-food-climate

p.23 ***John Lanchester's recent novel***: Lanchester (2019).

p.24 ***explored in scientific modelling***: most authoritatively, in the IPCC Report already cited (available at: https://www.ipcc.ch/sr15/download/).

p.28 ***I have written elsewhere about denial of this kind***: in Foster (2015), Part I *passim*.

p.28 ***'No, of course it's not too late'***: quoted in the Green Party magazine *Green World*, No. 84 (Spring 2014).

p.29 ***my fellow philosopher Rupert Read***: see Read (2021).

Chapter 2

p.32 ***the American writer Rebecca Solnit***: see Solnit (2004/2016), p.109.

pp.32–3 ***Joanna Macy and Chris Johnstone***: in Macy and Johnstone (2012), pp.189–91.

p.33 ***the Extinction Rebellion 'handbook'***: Extinction Rebellion (2019), pp.11,13.

p.35 ***this means around 3.5 per cent of the relevant population***: the work regularly cited in this connection is Chenoweth and Stephan (2011).

p.35 ***Naomi Klein and … George Monbiot***: see Klein (2014), Monbiot (2017).

p.35 ***as Klein in particular emphasizes***: Klein (2014) – see especially Chapter 10.

p.37 ***Active hope means accepting …***: see Macy and Johnstone (2012) – the quotations come respectively from pp.37 and 221.

p.37 ***as Solnit expresses the same idea***: Solnit (2004/2016), p.xiv.

p.37 ***'Say not the struggle naught availeth'***: see Clough ed. Mulhauser (1974), p.452. (The whole poem is closely relevant.)

p.37 ***'Wars will break out …'***: Solnit (2004/2016), pp.4–5.

p.37 ***David Hume famously pointed out***: in Hume (1777/1975), pp.25–6, 35–6.

p.38 ***Andrew Simms of the Rapid Transition Alliance***: https://www.bbc.com/ideas/videos/can-we-transform-the-world-in-12-years/p073j3z5

p.38 ***George Monbiot in his book* Out of the Wreckage**: Monbiot (2017). I should stress that, as in Foster (2015), I criticize Monbiot gratefully; he has the huge merit in a writer on these matters of being always clear and worth arguing with, and anyone interested in the human future should read his book.

p.39 ***Shelley at his shrillest***: see Shelley (1967), p.34 – 'The Mask of Anarchy' (written in response to Peterloo and addressed to the 'Men of England, heirs of Glory'), with its famously hopeful Stanza XXXVIII:

Rise like Lions after slumber
In unvanquishable number
… Ye are many, they are few.

p.40 ***'The great emancipations … came about through the mass mobilisation of citizens'***: Monbiot (2017), p.168.

p.40 ***an idea that can be adapted to any situation***: Monbiot (2017), pp.173–4.

p.41 ***'expect to be astonished'***: Solnit (2004/2016), p.109.

p.42 ***an initially rather surprising source … keeping to 2°C***: in Anderson (2012), p.38.

p.42 ***as Al Gore and many others remind us***: see for instance https://www.algore.com/about/the-climate-crisis

p.42 ***'… Anyone who gets on a plane once a year'***: Anderson (2012), p.36.

p.44 ***Faith, as Tolstoy says***: in Tolstoy (1921), p.60.

p.45 ***the language in which XR expresses its vision***: see https://extinctionrebellion.uk/the-truth/about-us/

Chapter 3

p.49 ***one misconceived psychological account***: see https://www.psychologytoday.com/us/blog/going-out-not-knowing/201909/hoping-against-hope

p.50 ***Churchill's capacity for hoping against hope***: see (and hear) https://winstonchurchill.org/resources/speeches/1940-the-finest-hour/we-shall-fight-on-the-beaches/

p.54 ***since Hume***: see Hume (1777/1975), especially the chapter 'Of Liberty and Necessity', pp.80ff.

p.55 ***Daniel Dennett's 'Cartesian theatre'***: see Dennett (1992), *passim*.

p.56 ***A particularly egregious example***: see Crick (1994), p.3. The single mention and immediate dismissal of subjectivity in this book comes on p.21.

p.56 ***Nor is Crick by any means alone***: see also Dennett (1992) and Metzinger (2004).

p.58 ***Britain's carbon emissions***: see data on this from the Union of Concerned Scientists at: https://www.ucsusa.org/resources/each-countrys-share-co2-emissions

p.58 ***'the tragedy of the commons'***: see Hardin (1968).

p.60 ***'No one is too small to make a difference'***: the title of Thunberg (2019).

p.60 ***climate change does not involve just single, intra- and intergenerational collective action problems***: see Jamieson (2014), p.100.

p.62 ***emissions exceeding the annual total of someone in Uganda or Somalia***: see, for instance, https://www.theguardian.com/environment/ng-interactive/2019/jul/19/carbon-calculator-how-taking-one-flight-emits-as-much-as-many-people-do-in-a-year

p.62 ***'The bigger your carbon footprint – the bigger your moral duty. …'***: Thunberg (2019), p.24.

Chapter 4

p.64 ***the 'golden rule'***: as Simon Blackburn notes in Blackburn (2001), p.101, this principle appears in one form or another in almost every ethical tradition.

p.67 ***an analogy due to the philosopher Derek Parfit***: from Parfit (1986), p.356.

p.67 ***the Harmless Torturers***: Parfit (1986), p.79.

p.68 ***some Kantian deontological requirement***: for both forms discussed, see Kant (1785/1964).

p.70 ***The carbon footprint of an individual or an activity***: for the definition, see for instance the Carbon Trust guide at: https://www.carbontrust.com/resources/carbon-footprinting-guide

p.70 ***an* ecological *footprint***: here the classic text is Wackernagel and Rees (1996).

p.71 ***'The best way to do this is to reduce your carbon footprint …'***: see https://www.goodenergy.co.uk/blog/2017/11/20/what-is-a-carbon-footprint/

p.71 ***A cheerful recent book***: Berners-Lee (2020) – actually a very useful guide to reducing one's emissions exposure, if one discounts the dubious metaphor.

p.75 ***E.F. Schumacher's old advice to do the right thing and not bother your head***: this is probably from either Schumacher (1974) or (1978), but I can't just at the moment track it down in either. It certainly comes up if you Google 'Schumacher quotes', for example at: https://www.inspiringquotes.us/author/4078-e-f-schumacher/about-soul

p.75 ***Lawrence's injunction to find and follow your deepest impulse***: see Lawrence (1964) – the Introduction and the chapter on Franklin are both highly relevant.

p.76 ***as Schopenhauer noted***: see Schopenhauer (1841/2009), p.264.

p.76 ***Freya Mathews … glossing the late Norwegian philosopher Arne Næss***: see Mathews (1991), p.149.

p.77 ***love, as Iris Murdoch wisely remarked***: somewhere in Murdoch (1999), though I have been unable to track down the page number.

p.82 ***St Mawr in Lawrence's tale of that name***: in Lawrence (1950) – the quotation is from pp.56–7.

p.82 ***'This civilisation is finished'***: the title and theme of Read and Alexander (2019).

Chapter 5

p.85 ***the aspiration 'to put everything right'***: Ernman and Thunberg (2020), p.275.

p.86 ***the terrain demarcated by … Hannah Arendt***: see Arendt (1958) – see Arendt (1958), pp.177–8.

p.86 ***Al Gore's more recent film***: 'An Inconvenient Sequel: Truth to Power', available at: https://wvw.heerofilms.com/en/432602-1569480880/2017

p.87 ***'Faith precedes all attempts to derive it from something else'***: see Tillich (1957), p.8.

p.87 ***'A map of the world that does not include Utopia …'***: from Wilde's 1891 essay 'The Soul of Man Under Socialism', in Wilde (1948), p.1059.

p.88 ***as Krishan Kumar glosses this***: in Kumar (1991), pp.95–6.

p.88 ***what Dr Johnson called 'the vanity of human wishes'***: see his 1749 poem of this title in Johnson (1968), pp.139–48. His anti-utopian admonition to:

> *… hope not life from grief or danger free,*
> *Nor think the doom of Man revers'd for thee*

might usefully be taken to heart as a guide to life in general.

p.88 ***'the crooked timber of humanity'***: the phrase is Kant's, from an essay of 1795 – see Kant (2006), p.9.

p.89 ***a world without weapons***: see Macy and Johnstone (2012), p.170.

p.93 ***Human beings are born to suffer***: see Hamilton (2016), p.9.

p.93 ***Hamilton ... quotes William James***: Hamilton (2016), p.87. The quotation itself is there cited as coming from James (1902/2008), p.65.

p.94 ***Mary Midgley notes ...*** : see Midgley (1979), *passim.*

p.95 ***As the philosopher John McDowell puts it***: in McDowell (1998), p.188.

p.97 ***the Vicar of Bray's successive submissions***: as expressed in the well-known song named after him, and concluding, in regard to the Hanoverian succession, with the classically pragmatic declaration that:

> *... in my Faith, and Loyalty,*
> *I never once will falter,*
> *But George, my lawful king shall be,*
> *Except the Times shou'd alter.*

p.98 **oikophilia *(Roger Scruton's useful coinage ...)***: see Scruton (2013), especially pp.326–75.

p.98 ***as ... Peter Singer eagerly puts it***: in Singer (2011), p.119.

p.99 ***closer to the etymological root***: which, says the *Shorter Oxford English Dictionary*, is Old English *aght*, from *agan*, to owe.

p.100 ***the 'metaphysical solace which ... we derive from every true tragedy ...'***: see Nietzsche (1872/1999), p.39.

p.100 ***Man was made for joy and woe***: see the poem 'Auguries of Innocence' from about 1803, in Blake (1946), p.119.

Chapter 6

p.104 ***a paradigm shift may be in the offing***: on such shifts, see Kuhn (1962), *passim.*

p.105 ***what have come to be called 'wicked problems'***: this notion goes back to a now famous paper by two planning academics from Berkeley – see Rittel and Webber (1973).

p.105 ***according to the sociologist of science Steve Rayner***: see his 'Rethinking Environmental Problems', lecture available at: https://www.youtube.com/watch?v=PEkvP3EUKJg

p.106 ***recent insightful criticism of the top-down global regime***: see for example G. Prins and S. Rayner, 'The Wrong Trousers: Radically Rethinking Climate Policy', available at: http://eureka.sbs.ox.ac.uk/66/

p.106 ***recent commentators have sought to distinguish out the problem of climate change***: see on this Levin et al (2012).

p.107 ***the 'contradictory certitudes' characterizing wicked problems***: Rayner in the lecture already referred to at p.105.

p.108 ***as advocated by Marco Verweij***: in Verweij (2011).

p.108 ***Mike Hulme***: see Hulme (2009), p.312.

p.111 ***the 'everyday facts of lived practical reason'***: see Nussbaum (2001), p.5.

p.114 ***the religion-substitute classically formulated … by John Stuart Mill***: see his 1850s essay on 'The Utility of Religion' in Mill (1985), p.420.

p.115 ***Barry Commoner's still valuable book***: Commoner (1972).

p.115 ***James Lovelock's rich metaphor***: as persuasively deployed in a series of books from Lovelock (2000) to Lovelock (2006).

p.116 ***'… putting wisdom in place of science as the highest goal'***: Nietzsche (1872/1999), p.87.

p.118 ***The humans that come after are not going to give a shit about whether we were pacifists …***: see the interview with Jensen in *Dark Mountain* (2010), p.110.

Chapter 7

p.121 ***'Climate change … is domination'***: Holthaus (2020), p.198.

p.121 ***Jonathon Porritt, on why governments failed***: see Porritt (2020), p.217.

p.121 ***an online paper which has apparently set records***: 'Deep Adaptation: A Map for Navigating Climate Tragedy', IFLAS Occasional Paper 2 (July 2018, revised 2020), available at: http://www.lifeworth.com/deepadaptation.pdf

p.122 ***climate perturbations held to be … in principle manageable***: 'Deep Adaptation', p.6.

p.122 ***the Stockholm Resilience Centre***: see their 'What is Resilience?', available at: https://www.stockholmresilience.org/research/research-news/2015-02-19-what-is-resilience.html

p.122 ***'… "bouncing back" from difficult experiences'***: 'Deep Adaptation', p.22. The quotation is from 'Building Your Resilience', American Psychological Association (2018), available at: https://www.apa.org/topics/resilience

p.122 ***Bendell's gloss on such rebounding***: 'Deep Adaptation', p.22. His comments on *relinquishment* and *restoration* (see p.124) are also to be found on the same page.

p.125 ***'the remaining 2°C budget demands* revolutionary change*'***: see Klein (2019), p.117. The original text from which she quotes is available at: http://kevinanderson.info/blog/why-carbon-prices-cant-deliver-the-2c-target/

p.126 ***Environmentalists can't win the emissions-reduction fights on our own***: Klein (2019), pp.201–2.

p.127 ***caring about the end of the world and the end of the month***: Klein (2019), pp.287–8.

p.127 ***Klein's LEAP Manifesto***: see Klein (2017), especially Chapter 13 and the Postscript.

p.127 ***key values in this context***: Klein (2017), p.267.

p.128 ***governments ... can stand 'moral force' till the cows come home***: see the 1942 essay 'Pacifism and the War' in Orwell (1970a), Vol 2, p.262.

p.128 ***'... democratically and without a bloodbath'***: see Klein (2014), p.452.

p.128–9 ***'The populace at large does not rebel for fun ...'***: Dunn (1989), p.246.

p.129 **Common Sense for the 21st Century**: available at https://www.rogerhallam.com/wp-content/uploads/2019/08/Common-Sense-for-the-21st-Century_by-Roger-Hallam-Download-version.pdf

p.129 ***Let's be frank about what 'catastrophe' actually means ...***: *Common Sense for the 21st Century*, p.14.

p.130 ***only 3.5 per cent of the population needs to get drawn into these activities***: this is the finding of Chenoweth and Stephan (2011), which Hallam cites at p.30. The 3.5 per cent figure is in fact rather buried within the statistical apparatus of this book (which also does *not* conclude that *only* non-violent campaigns work), but is emphasized as a headline by one of the authors at, for example, https://www.nonviolent-conflict.org/resource/success-nonviolent-civil-resistance/

p.130 ***the campaign's demands***: for which, see its website at: https://extinctionrebellion.uk/the-truth/demands/

p.131 ***the last thirty years of abject failure***: *Common Sense for the 21st Century*, p.6.

p.131 ***'exposed to 360° understanding of an issue'***: *Common Sense for the 21st Century*, p.63.

p.132 ***The examples which tend to get cited***: as, for instance, in the *Extinction Rebellion Guide to Citizens' Assemblies* (2019), available at: https://extinctionrebellion.uk/wp-content/uploads/2019/06/The-Extinction-Rebellion-Guide-to-Citizens-Assemblies-Version-1.1-25-June-2019.pdf, pp.18–25.

p.132 ***as Herbert Marcuse once put it***: in Marcuse (1964/2002), p.11.

p.133 ***the UK Climate Citizens' Assembly***: for all these details, see its Report at https://www.climateassembly.uk/report/

p.134 ***'... deliberation and reason will finally be given space ...'***: *Common Sense for the 21st Century*, p.62.

Chapter 8

p.137 ***Marcuse's enthusiastic welcome for the student movement***: in Marcuse (1969), p.67.

p.137 ***'… a new version of the children's crusade …'***: MacIntyre (1970), p.89.

p.138 ***Conscious righteousness riding on instinctual revolt***: Marcuse (1969), p.24.

p.140 ***all the way with Nietzsche***: as in, classically, Nietzsche (1886/1973) and (1887/1996).

p.141 ***Marx … remained so ambiguous***: on this see Cohen (1983), who argues persuasively that Marx didn't altogether know his own mind here.

p.144 ***our use of fossil fuels is a Faustian pact***: see Monbiot (2006), especially Chapter 1.

p.145 ***life-meaning is … a profound human need***: see Frankl (1946/2004).

p.148 ***Hearts with one purpose alone***: from 'Easter 1916', in Yeats (1967), p.204.

p.149 ***'If the emissions have to stop, then we must stop the emissions …'***: Thunberg (2019), p.7.

p.149 ***'a moment at which humanity summons enough courage …'***: see Foster (ed.) (2019), p.12.

p.152 ***'It is clear why we are here in this world …'***: Hallam's *Common Sense for the 21st Century*, p.66.

Chapter 9

p.156 ***'We need to focus every inch of our being on climate change …'***: Thunberg (2019), p.37.

p.158 ***'democracy's ugly sister'***: I came across this neat but misleading description in a letter to *The Guardian* from (I assume, pure nominal coincidence aside) the political thinker David Marquand, published 26 October 2020. (I am in time now to add that some 73 million people voted in November 2020 for Trump, even though he didn't win. Trump is a cancer in the US body politic, but he is not an *accident*, any more than the cancer which you get from a lifetime's smoking habit is an accident.)

p.159 ***'All men are created equal'***: this is, of course, how it is put in the American Declaration of Independence, but Jefferson got his ideas on these matters, via Lafayette, Voltaire and others, out of Locke, especially Locke (1689/2008).

p.159 ***every head to count for one, in Bentham's blunt idiom***: This version of the principle is cited and called 'Bentham's dictum' by J.S. Mill in *Utilitarianism* – see Mill (1863/1998), p.149. Bentham's own wording is 'every individual in the country tells for one; no individual for more than one': Bentham (1827), iv 475.

p.160 ***'even quite common men have souls'***: Tawney (1938), p.278.

p.161 ***ecofascism as defined by the environmental historian Michael Zimmerman***: see his 'Ecofascism', in Taylor (2008), pp.531–2.

p.161 ***Churchill, for instance, reckoned …***: see Churchill (1949), p.315.

p.163 ***the technical forestry concept of* Nachhaltigkeit**: see on this Ulrich Grober, *Die Entdeckung der Nachhaltigkeit* (Munich: Verlag Antje Kunstmann, 2010), translated into English as Grober (2012).

p.165 ***Rainborough's ringing words***: as quoted in Trevelyan (1904), p.282.

p.169 ***'deep green resistance'***: see the book of that title by McBay et al (2011), also available via https://deepgreenresistance.org/about-us/

p.170 ***subsidiarity created from the bottom up***: see, for instance, Helena Norberg-Hodge and Rupert Read, *Post-growth Localisation* (Local Futures and Green House, 2016), available at: https://www.greenhousethinktank.org/uploads/4/8/3/2/48324387/post-growth-localisation_pamphlet.pdf

p.172 ***According to the World Health Organization***: see on this https://www.wired.co.uk/article/china-coronavirus. Also Malm (2020), especially Chapter 2.

p.172 ***readier to bring these and similar lifestyle changes to bear***: see, for instance, the very interesting discussion of these issues sponsored by Green House, available at: https://www.greenhousethinktank.org/past-events.html

p.174 ***The foreword to the Party's Political Programme***: available at https://www.greenparty.org.uk/assets/images/national-site/political-programme-web-v1.3.pdf

p.174 ***a book-length portfolio***: for access to which, see https://policy.greenparty.org.uk/

p.175 ***making 'the desirable feasible'***: see the Political Programme, just referenced, p.2.

References

Anderson, K. (2012) 'Climate change going beyond dangerous: brutal numbers and tenuous hope', *Development Dialogue*, September 2012: 16–40.

Arendt, H. (1958) *The Human Condition*, Chicago, IL: University of Chicago Press.

Attenborough, D. (2020) *A Life on Our Planet*, London: Witness Books.

Bentham, J. (1827) J.S. Mill (ed) *Rationale of Judicial Evidence, Specially Applied to English Practice*, London: Hunt and Clarke.

Berners-Lee, M. (2020) *How Bad Are Bananas? The Carbon Footprint of Everything*, London: Profile Books.

Berners-Lee, M. and Clark, D. (2013) Foreword by B. McKibben, *The Burning Question: We Can't Burn Half the World's Oil, Coal and Gas. So How Do We Quit?*, London: Profile.

Blackburn, S. (2001) *Ethics: A Very Short Introduction*, Oxford: Oxford University Press.

Blake, W. (1946) G. Keynes (ed) *Poetry and Prose of William Blake*, London: Nonesuch.

Chenoweth, E. and Stephan, M. (2011) *Why Civil Resistance Works: The Strategic Logic of Non-Violent Conflict*, New York, NY: Columbia University Press.

Churchill, W. (1949) *The Second World War*, Vol 2, London: Cassell.

Clough. A. (1974) F.L. Mulhauser (ed) *The Poems of Arthur Hugh Clough*, Oxford: Oxford University Press.

Cohen, G. (1983) 'Review of Allen Wood, *Karl Marx*', *Mind*, 92: 440–5.

Commoner, B. (1972) *The Closing Circle: Confronting the Environmental Crisis*, London: Cape.

Crick, F. (1994) *The Astonishing Hypothesis: The Scientific Search for the Soul*, New York, NY: Touchstone.

Dark Mountain (2010) *Dark Mountain*, Issue 1 (Summer), Bodmin and King's Lynn: MPG Books.

Dennett, D. (1992) *Consciousness Explained*, London: Allen Lane.

Dunn, J. (1989) *Modern Revolutions: An Introduction to the Analysis of a Political Phenomenon*, 2nd edn, Cambridge: Cambridge University Press.

Ernman M., Ernman, B., Thunberg, G. and Thunberg, S. (2020) trans. P. Norlen and S. Vogel, *Our House is On Fire: Scenes of a Family and a Planet in Crisis*, London: Allen Lane.
Extinction Rebellion (2019) *This Is Not a Drill*, London: Penguin.
Foster, J. (ed) (1997) *Valuing Nature?*, London: Routledge.
Foster, J. (2008) *The Sustainability Mirage*, London: Earthscan.
Foster, J. (2015) *After Sustainability*, Abingdon: Routledge.
Foster, J. (ed) (2019) *Facing Up to Climate Reality: Honesty, Disaster and Hope*, London: Green House.
Frankl, V. (1946/2004) *Man's Search for Meaning*, London: Rider.
Geuss, R. (2008) *Philosophy and Real Politics*, Princeton, NJ: Princeton University Press.
Grober, U. (2012) trans. R. Cunningham, *Sustainability – A Cultural History*, Totnes: Green Books.
Hamilton, Ch. (2016) *A Philosophy of Tragedy*, London: Reaktion.
Hamilton, Cl. (2010) *Requiem for a Species*, London: Earthscan.
Hamilton, Cl. (2017) *Defiant Earth: The Fate of Humans in the Anthropocene*, Cambridge: Polity Press.
Hardin, G. (1968) 'The Tragedy of the Commons', *Science*, 162(3859): 1243–8.
Helm, D. (2012) *The Carbon Crunch*, New Haven, CT and London: Yale University Press.
Holthaus, E. (2020) *The Future Earth*, New York, NY: HarperCollins.
Hulme, M. (2009) *Why We Disagree about Climate Change*, Cambridge: Cambridge University Press.
Hume, D. (1777/1975) L. Selby-Bigge and P. Nidditch (eds) *Enquiries Concerning Human Understanding and Concerning the Principles of Morals*, Oxford: Clarendon Press.
Jackson, T. (2009) *Prosperity Without Growth*, London: Earthscan.
James, W. (1902/2008) *The Varieties of Religious Experience*, Rockville, MD.
Jamieson, D. (2014) *Reason in a Dark Time*, Oxford: Oxford University Press.
Johnson, S. (1968) P. Crutwell (ed) *Samuel Johnson: Selected Writings*, Harmondsworth: Penguin.
Kant, I. (1785/1964) trans. H. Paton (ed) *Groundwork of the Metaphysic of Morals*, New York, NY: Harper and Row.
Kant, I. (2006) trans. D. Colclasure, P. Kleingeld (ed) *Towards Perpetual Peace and Other Writings*, New Haven, CT: Yale University Press.
Klein, N. (2014) *This Changes Everything: Capitalism vs. the Climate*, Harmondsworth: Penguin.
Klein, N. (2017) *No Is Not Enough*, London: Penguin.
Klein, N. (2019) *On Fire: The Burning Case for a Green New Deal*, London: Allen Lane.
Kuhn, T. (1962) *The Structure of Scientific Revolutions*, Chicago, IL: University of Chicago Press.

Kumar, K. (1991) *Utopianism*, Buckingham: Open University Press.

Lanchester, J. (2019) *The Wall*, London: Faber & Faber.

Lawrence, D. (1950) *St Mawr/The Virgin and the Gipsy*, Harmondsworth: Penguin.

Lawrence, D. (1964) *Studies in Classic American Literature*, London: Heinemann.

Levin, K., Cashore, B., Bernstein, S. and Auld, G. (2012) 'Overcoming the tragedy of super wicked problems: constraining our future selves to ameliorate global climate change', *Policy Science*, 45: 123–52.

Locke, J. (1689/2008) I. Shapiro (ed) *Two Treatises of Government*, New Haven, CT: Yale University Press.

Lovelock, J. (2000) *Gaia: A New Look at Life on Earth*, Oxford: Oxford University Press.

Lovelock, J. (2006) *The Revenge of Gaia*, London: Penguin.

MacIntyre, A. (1970) *Marcuse*, London: Fontana.

Macy, J. and Johnstone, C. (2012) *Active Hope*, Novato, CA: New World Library.

Malm, A. (2020) *Corona, Climate, Chronic Emergency*, London: Verso.

Marcuse, H. (1964/2002) *One-Dimensional Man*, London: Routledge.

Marcuse, H. (1969) *An Essay on Liberation*, Harmondsworth: Penguin.

Martin, A. (2013) *How We Hope: A Moral Psychology*, Princeton, NJ: Princeton University Press.

Mathews, F. (1991) *The Ecological Self*, London: Routledge.

McBay A., Keith, L. and Jensen, D. (2011) *Deep Green Resistance*, New York, NY: Seven Stories Press.

McCarthy, C. (2006) *The Road*, London: Picador.

McDowell, J. (1998) *Mind, Value and Reality*, London: Harvard University Press.

Metzinger, T. (2004) *Being No One: The Self-Model Theory of Subjectivity*, Cambridge, MA: MIT Press.

Midgley, M. (1979) *Beast and Man: The Roots of Human Nature*, Hassocks: Harvester Press.

Mill, J.S. (1985) J.M. Robson (ed) *The Collected Works of John Stuart Mill, Vol X: Essays on Ethics, Religion and Society*, Toronto and London: University of Toronto Press/Routledge and Kegan Paul.

Mill, J.S. (1863/1998) R. Crisp (ed) *Utilitarianism*, Oxford: Oxford University Press.

Monbiot, G. (2006) *Heat: How We Can Stop the Planet Burning*, London: Allen Lane.

Monbiot, G. (2017) *Out of the Wreckage*, London: Verso.

Morris, W. (1891) *News from Nowhere, or, An Epoch of Rest*, London: Reeves and Turner.

Mulgan, T. (2011) *Ethics for a Broken World*, Durham: Acumen.

Murdoch, I. (1999) P. Conradi (ed) *Existentialists and Mystics: Writings on Philosophy and Literature*, Harmondsworth: Penguin.
Nagel, T. (1989) *The View From Nowhere*, Oxford: Oxford University Press.
Nietzsche, F. (1872/1999) trans. R. Speirs, R. Geuss and R. Speirs (eds) *The Birth of Tragedy and Other Writings*, Cambridge: Cambridge University Press.
Nietzsche, F. (1886/1973) trans. R. Hollingdale, *Beyond Good and Evil*, London: Penguin.
Nietzsche, F. (1887/1996) trans. D. Smith, *On the Genealogy of Morals*, Oxford: Oxford University Press.
Nussbaum, M. (2001) *The Fragility of Goodness*, Cambridge: Cambridge University Press.
Orwell, G. (1970a) S. Orwell and I. Angus (eds) *The Collected Essays, Journalism and Letters of George Orwell*, Vol 2, Harmondsworth: Penguin.
Orwell, G. (1970b) S. Orwell and I. Angus (eds) *The Collected Essays, Journalism and Letters of George Orwell*, Vol 4, Harmondsworth: Penguin.
Parfit, D. (1986) *Reasons and Persons*, Oxford: Oxford University Press.
Porritt, J. (2020) *Hope in Hell*, London: Simon & Schuster.
Read, R. (2021) *Parents for a Future: How Loving Our Children Can Prevent Climate Collapse*, Norwich: UEA Publishing Project.
Read, R. and Alexander, S. (2019) *This Civilisation is Finished*, Melbourne: Simplicity Institute.
Rittel, H. and Webber, M. (1973) 'Dilemmas in a general theory of planning', *Policy Sciences*, 4: 155–69.
Schopenhauer, A. (1841/2009) trans. D. Cartwright and E. Erdmann, *The Two Fundamental Problems of Ethics*, Oxford: Oxford University Press.
Schumacher, E. (1974) *Small Is Beautiful*, London: Abacus.
Schumacher, E. (1978) *A Guide for the Perplexed*, London: Abacus.
Scruton, R. (2013) *Green Philosophy: How to Think Seriously about the Planet*, London: Atlantic Books.
Shelley, P. (1967) T. Hutchinson (ed) *Poetical Works*, London: Oxford University Press.
Singer, P. (2011) *The Expanding Circle: Ethics, Evolution and Moral Progress*, 2nd edn, Princeton, NJ: Princeton University Press.
Solnit, R. (2004/2016) *Hope in the Dark: Untold Histories, Wild Possibilities*, Chicago, IL: Haymarket.
Tawney, R. (1938) *Religion and the Rise of Capitalism*, Harmondsworth: Penguin.
Taylor, B. (ed) (2008) *Encyclopedia of Religion and Nature*, Vol 1, London: Continuum.
Thunberg, G. (2019) *No One Is Too Small to Make a Difference*, London: Penguin.
Tillich, P. (1957) *Dynamics of Faith*, New York, NY: Harper & Row.
Tolstoy, L. (1921) trans. A. Maude, *A Confession and What I Believe*, London: Oxford University Press.

Trevelyan, G. (1904) *England under the Stuarts*, London: Methuen.
Verveij, M. (2011) *Clumsy Solutions for a Wicked World*, Basingstoke: Palgrave Macmillan.
Wackernagel, M. and Rees, W. (1996) *Our Ecological Footprint: Reducing Human Impact on the Planet*, Gabriola Island, BC: New Society Publishers.
Wallace-Wells, D. (2019) *The Uninhabitable Earth*, London: Allen Lane.
Wilde, O. (1948) J.B. Foreman (ed) *The Complete Works of Oscar Wilde*, London: Collins.
Yeats, W. (1967) *The Collected Poems of W.B. Yeats*, London: Macmillan.

Index

References to figures and photographs appear in *italic* type.